Viva Studymates

Chemistry

Explanations and Practice

Viva Studymates

Chemistry

Explanations and Practice

Dr Aleyamma Ninan

Viva Books

New Delhi | Mumbai | Chennai | Kolkata | Bangalore | Hyderabad | Kochi | Guwahati

First Indian Edition 2010

VIVA BOOKS PRIVATE LIMITED

- 4737/23 Ansari Road, Daryaganj, New Delhi 110 002
  E-mail: vivadelhi@vivagroupindia.net, Tel. 42242200
- 76, Service Industries, Shirvane, Sector 1, Nerul, Navi Mumbai 400 706
  E-mail: vivamumbai@vivagroupindia.net, Tel. 27721273, 27721274
- Jamals Fazal Chambers, 26 Greams Road, Chennai 600 006
  E-mail: vivachennai@vivagroupindia.net, Tel. 28290304, 28294241
- B-103, Jindal Towers, 21/1A/3 Darga Road, Kolkata 700 017
  E-mail: vivakolkata@vivagroupindia.net, Tel. 22836381, 22816713
- 7, Sovereign Park Apartments, 56-58, K. R. Road, Basavanagudi, Bangalore 560 004
  E-mail: vivabangalore@vivagroupindia.net, Tel. 26607409, 26607410
- 101-102 Mughal Marc Apartments, 3-4-637 to 641, Narayanguda, Hyderabad 500 029
  E-mail: vivahyderabad@vivagroupindia.net, Tel. 27564481, 27564482
- Beevi Towers, First Floor, SRM Road, Kaloor, Kochi 682 018, Kerala
  E-mail: vivakochi@vivagroupindia.net, Tel: 0484-2403055, 2403056
- 232, GNB Road, Beside UCO Bank, Silpukhuri, Guwahati 781 003
  E-mail: vivaguwahati@vivagroupindia.net, Tel: 0361-2666386

www.vivagroupindia.com

Published by arrangement with
Studymates Limited
PO Box 2, Bishops Lydeard,
Somerset TA4 3YE,
United Kingdom

ISBN : 978-81-309-1304-9

Published by Vinod Vasishtha for Viva Books Private Limited, 4737/23, Ansari Road, Daryaganj, New Delhi-110 002. Printed and bound by Swan Press, New Delhi.

G0000 / G0600

Contents

Preface

Drawn from my extensive teaching experience, I have crafted this book to provide clear explanations of underlying concepts yet succinct enough to be used as a study guide. It is intended to supplement, not replace the students' main texts. It has also been designed as a novel, incisive way of presenting important topics and is primarily aimed at increasing understanding of essential concepts. The nine chapters include 'Atomic Structure and Periodic Table', 'Formulae, Equations and Stoichiometry', 'Energy Changes', 'Chemical Kinetics' and 'Acids and Bases'. Each chapter starts with a 'one minute summary' followed by the main text which is lucidly presented in a logical and clear note form, liberally sprinkled with illustrations and examples. Emphasis is given to clearly set out examples and calculations in all chapters, again with the purpose of improving comprehension. 'Practice questions' and 'progress questions' complete each chapter with answers to these questions found at the end of the book.

I would like to thank Dr. David Brown and Dr. Golda Ninan for reading the entire text and giving suggestions. Finally I would like to thank my family for their support and encouragement throughout the period of writing this book.

A. Ninan

1 Atomic Structure and Periodic Table

One-minute summary

Atoms are made up of three fundamental particles; protons, electrons and neutrons. The arrangement of these particles in atoms and their charges and masses form the main theme of this chapter. The arrangement of electrons in orbits or energy levels is discussed in detail. The 110 elements known to scientists are classified in the form of a table called the Periodic Table. The basis for the arrangement of elements in the Periodic Table is atòmic number and electronic configuration and this together with the periodic properties are discussed in this chapter.

The highlights of this chapter are

- the masses and charges of protons, electrons and neutrons
- atomic number and mass number
- isotopes
- electronic configurations
- shape of s and p orbitals
- the Periodic Table
- periodic properties

1.1 Fundamental particles – What is an atom made of?

An **atom** consists of three fundamental particles: protons, neutrons and electrons. A **proton** is a positively charged particle with a mass of 1 unit (1u, formerly 1 amu) in the **atomic mass unit scale**. A **neutron** is neutral and the mass of a neutron is also 1u. An **electron** is a negatively charged particle the mass of which is approximately 1/1850 the mass of a proton. These are summarised below (Table 1.1).

Table 1.1: Properties of the fundamental particles

Fundamental particle	Charge	Mass (u)
Proton	+1	1
Neutron	0	1
Electron	–1	negligible

Note: The mass of an atom is too small to express in grams. So, atomic mass unit scale is used. 1 u is exactly 1/12 the mass of a carbon-12 atom. In grams, 1 u = 1.6605×10^{-24} g.

The protons and neutrons are packed together in a compact mass and this forms the **nucleus** of the atom. The electrons are arranged around the nucleus.

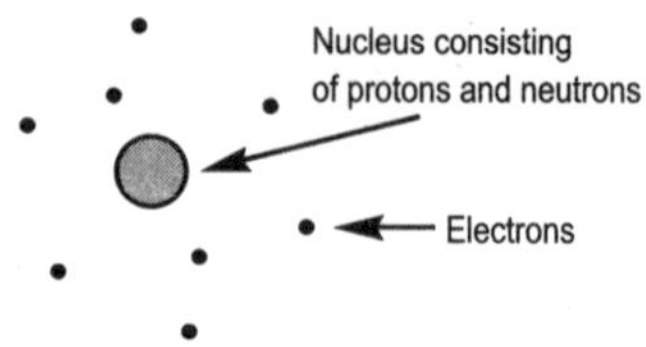

Figure 1.1: A simple representation of an atom

1.2 Atomic number, mass number and isotopes

An atom of any element contains equal numbers of protons and electrons. So an atom as a whole is neutral. The **atomic number** (Z) of an element is the number of protons or electrons in an atom and this number is characteristic for an element. The **mass number** (A) is the sum of protons and neutrons. It is common practice to write the atomic number as a subscript and the mass number as a superscript as shown below.

$$^{1}_{1}H \qquad ^{7}_{3}Li \qquad ^{23}_{11}Na$$

This representation shows that the atomic number of hydrogen, H, is 1 and its mass number 1. The atomic number of lithium, Li, is 3 and its mass number is 7, and the atomic number of sodium, Na, is 11 and its mass number 23 (Table 1.2).

Isotopes are atoms of the same element with equal numbers of protons but a different number of neutrons, hence different masses. We can take two of the isotopes of carbon, ^{12}C and ^{13}C, as example. Both isotopes contain six protons and six electrons each. But the nucleus of ^{12}C contains six neutrons and that of ^{13}C seven neutrons (Table 1.2). Hydrogen forms three isotopes: ^{1}H, ^{2}H and ^{3}H. All three isotopes contain one proton and one electron each. ^{1}H has no neutron, ^{2}H has one neutron and ^{3}H has two.

Table 1.2

Element	Atomic No. Z	Mass No. A	No. of protons = Z	No. of electrons = Z	No. of neutrons = $A–Z$
$^{1}_{1}H$	1	1	1	1	0
$^{7}_{3}Li$	3	7	3	3	4
$^{12}_{6}C$	6	12	6	6	6
$^{13}_{6}C$	6	13	6	6	7
$^{23}_{11}Na$	11	23	11	11	12

Practice question 1.1 Write the numbers of protons, electrons and neutrons in these elements:
(a) He ($Z = 2$, $A = 4$) (b) Be ($Z = 4$, $A = 9$) (c) B ($Z = 5$, $A = 11$)

1.3 Electronic configuration: energy levels

Electrons are arranged around the nucleus in an atom at various *levels*. These levels are identified by the number n where **n** = 1, 2, 3 etc. starting from the innermost level.

These levels are called **shells** or **orbits**. They are also called **energy levels** since the energy of an electron in a level is quantized. This means that each electron possesses a discrete (fixed) amount of energy. An electron in a lower energy level has lower energy than one in a higher level. A given electron requires a certain, fixed amount of energy before it can move to a higher level. Similarly if an electron falls to a lower energy level, a fixed amount of energy is given out.

An **atomic orbital** or **energy level** is the region of space where an electron is found. Picture each negatively charged electron in an atom moving, vibrating and spinning constantly in *a region around the nucleus* forming a negatively charged electron cloud. The shape of the orbital is the shape of this cloud. The atomic orbital represents the region where the probability of finding a particular electron is the highest.

There are different types of atomic orbitals designated by the letters *s*, *p*, *d* and *f*. In the first energy level (n = 1) there is only one that being an *s* orbital (Figure 1.2 shows the shape of an s orbital). In the second energy level (n = 2) there are two types of orbitals, i.e., *s* and *p* (Figure 1.3), in the third level (n = 3), three types, i.e., *s*, *p* and *d* and in the fourth level (n = 4), four types, i.e., *s*, *p*, *d* and *f*.

There is only one *s* orbital in any energy level, that is, there is one *s* orbital in the first energy level, one *s* orbital in the second energy level and so on. There are three *p* orbitals in an energy level starting at the second energy level, five *d* orbitals starting at the third level and seven *f* orbitals starting at the fourth level. The *s* orbital of the first energy level is called a 1*s* orbital, a *p* orbital of the second energy level 2*p*, a *p* orbital of the third energy level 3*p* and so on. An orbital holds a maximum of two electrons and these two electrons spin in opposite directions. So an s orbital contains two electrons, the three *p* orbitals of an energy level six electrons and so on. These are summarised in Table 1.3.

Table 1.3

Energy level n	Number of the types of orbitals	No. and name of the orbitals	Number of electrons	Total number of electrons
1	one	one 1*s*	2	2
2	two	one 2*s* three 2*p*	2 6	8
3	three	one 3*s* three 3*p* five 3*d*	2 6 10	18
4	four	one 4*s* three 4*p* five 4*d* seven 4*f*	2 6 10 14	32

1.4 Shapes of s and p orbitals

An *s* orbital is spherical in shape and can be represented by a sphere as shown below. A 2*s* orbital is larger than a 1s orbital. For simplicity, a circle is drawn to represent an *s* orbital (Figure 1.2).

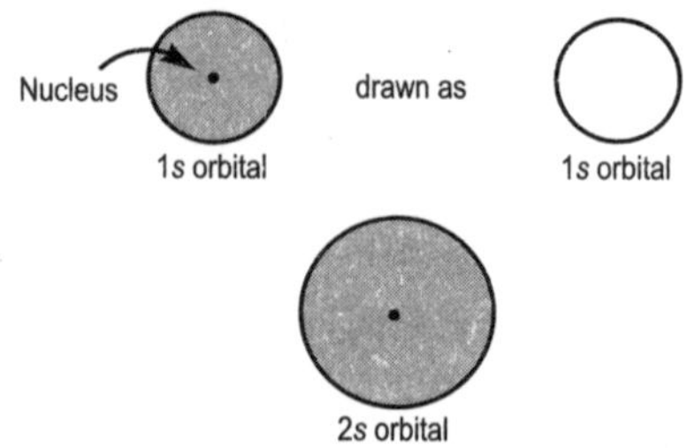

Figure 1.2: Shape of 1s and 2s orbitals

The *p* orbital is dumb-bell shaped and consists of two lobes, with the nucleus of the atom located between the two lobes. In any energy level, except in the first, there are three *p* orbitals of equal energy, designated p_x, p_y and p_z. They are directed along x, y and z axes and are perpendicular to one another (Figure 1.3).

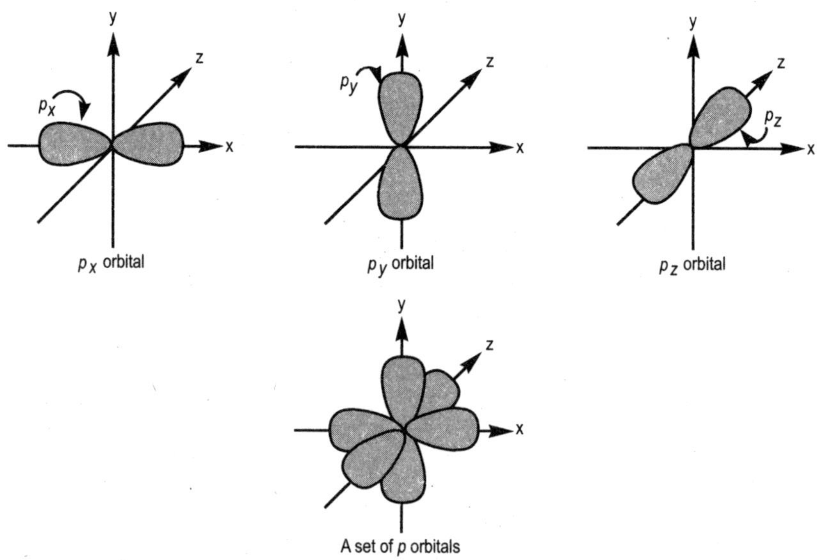

Figure 1.3: *p* orbitals

1.5 Electronic configuration

Electronic configuration is the arrangement of the electrons of an atom, in named atomic orbitals. The following rules are helpful in writing the electronic configurations of elements.

1. In an atom, electrons occupy the lowest energy level first, filling it before going to a higher level. This is called the **aufbau** or **building up principle.** The 1*s* orbital has the lowest energy value and is filled first, followed by 2*s*, then 2*p*, 3*s* and 3*p*.
2. An orbital is occupied by a maximum of two electrons (**Pauli exclusion principle**) and these two electrons spin in opposite directions.
3. When there is more than one orbital of equal energy value (for example, the three *p* orbitals of an energy level), electrons occupy them singly, before pairing takes place (**Hund's rule of maximum multiplicity**).

Now we can write the electronic configuration of some elements. A hydrogen atom (Atomic number, $Z = 1$) has one electron which occupies the 1s orbital. Its electronic configuration is written as:

H: $1s^1$
(Note that the superscript 1 stands for the number of electrons in the orbital.)

In the next element helium, He, ($Z=2$) there are two electrons and both these electrons occupy the $1s$ orbital.

He: $1s^2$

In the third element lithium, Li, ($Z = 3$), its 1s orbital is complete with two electrons, the third electron goes to the $2s$ orbital and its electronic configuration is,

Li: $1s^2 2s^1$

Example 1.1 Write the electronic configurations of beryllium, Be, ($Z = 4$), boron, B, ($Z = 5$) and carbon, C, ($Z = 6$).

Answer
Be: $1s^2\ 2s^2$
B: $1s^2\ 2s^2\ 2p^1$
C: $1s^2\ 2s^2\ 2p^2$

Note that the two p electrons of carbon are distributed in two of the three p orbitals (Hund's rule). The electronic configuration of carbon can be written as C: $1s^2 2s^2 2p_x^1 2p_y^1$ which further describes the electron distribution in the p orbitals.

Example 1.2 Write the electronic configuration of elements 7–10 in the Periodic Table.

Answer
Nitrogen, N ($Z = 7$): $1s^2\ 2s^2\ 2p_x^1\ 2p_y^1\ 2p_z^1$
Oxygen, O ($Z = 8$): $1s^2\ 2s^2\ 2p_x^2\ 2p_y^1\ 2p_z^1$
Fluorine, F ($Z = 9$): $1s^2\ 2s^2\ 2p_x^2\ 2p_y^2\ 2p_z^1$
Neon, Ne ($Z = 10$): $1s^2\ 2s^2\ 2p^6$

Note that when all the p orbitals of an energy level are filled, the arrangement can be written as p^6 instead of $p_x^2\ p_y^2\ p_z^2$.

The second energy level is complete with neon, Ne. In the next element sodium, Na, ($Z = 11$), one electron goes to the $3s$ orbital. The electronic configuration of sodium is,

Na: $1s^2\ 2s^2\ 2p^6\ 3s^1$ or, [Ne] $3s^1$
(i.e., the electronic configuration of Ne and $3s^1$.)

$3s$ and $3p$ orbitals are being filled in the elements magnesium, Mg, ($Z = 12$) to argon, Ar, ($Z = 18$). In the next two elements potassium, K, ($Z = 19$) and calcium, Ca, ($Z = 20$), electrons are added in the $4s$ orbital leaving the $3d$ orbitals vacant. This is because the energy needed to fill the $4s$ orbital is lower than that for the $3d$ orbital. In scandium, Sc, ($Z = 21$), one electron goes to the $3d$ orbital. The $3d$ orbitals are being filled in the next ten elements scandium, Sc, ($Z = 21$) to zinc, Zn, ($Z = 30$) (Table 1.4). Note the electronic configurations of the elements Ar, K, Ca, Sc and Zn.

Ar ($Z = 18$): [Ne] $3s^2 3p^6$
K ($Z = 19$): [Ar] $4s^1$
Ca ($Z = 20$): [Ar] $4s^2$
Sc ($Z = 21$): [Ar] $4s^2 3d^1$
Zn ($Z = 30$): [Ar] $4s^2 3d^{10}$

Order of filling orbitals

As we have seen in the above section, the order of filling orbitals is not straight forward. It can be clearly shown using this chart.

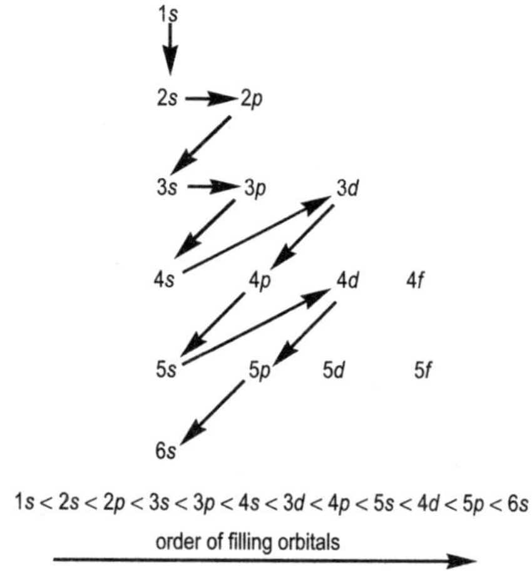

Figure 1.4: Order of filling orbitals

Practice question 1.2 Which atomic orbital in each of the following pairs of orbitals has a higher value of energy?
(a) 1*s* or 2*s* (b) 2*s* or 2*p* (c) 2*p* or 3*p* (d) 3*p* or 4*s* (e) 3*d* or 4*s*

The electronic structures of the first 36 elements are given below in Table 1.4.

Table 1.4: The electronic configurations of the first 36 elements

Element	Atomic number, Z	Electronic configuration	Element	Atomic number, Z	Electronic configuration
H	1	$1s^1$	N	7	[He]$2s^2 2p^3$
He	2	$1s^2$	O	8	[He]$2s^2 2p^4$
Li	3	[He]$2s^1$	F	9	[He]$2s^2 2p^5$
Be	4	[He]$2s^2$	Ne	10	[He]$2s^2 2p^6$
B	5	[He]$2s^2 2p^1$	Na	11	[Ne]$3s^1$
C	6	[He]$2s^2 2p^2$	Mg	12	[Ne]$3s^2$

(continued)

Element	Atomic number, Z	Electronic configuration	Element	Atomic number, Z	Electronic configuration
Al	13	[Ne]$3s^2\ 3p^1$	Mn	25	[Ar]$3d^5\ 4s^2$
Si	14	[Ne]$3s^2\ 3p^2$	Fe	26	[Ar]$3d^6\ 4s^2$
P	15	[Ne]$3s^2\ 3p^3$	Co	27	[Ar]$3d^7\ 4s^2$
S	16	[Ne]$3s^2\ 3p^4$	Ni	28	[Ar]$3d^8\ 4s^2$
Cl	17	[Ne]$3s^2\ 3p^5$	Cu	29	[Ar]$3d^{10}\ 4s^1$
Ar	18	[Ne]$3s^2\ 3p^6$	Zn	30	[Ar]$3d^{10}\ 4s^2$
K	19	[Ar]$4s^1$	Ga	31	[Ar]$3d^{10}\ 4s^2\ 4p^1$
Ca	20	[Ar]$4s^2$	Ge	32	[Ar]$3d^{10}\ 4s^2\ 4p^2$
Sc	21	[Ar]$3d^1\ 4s^2$	As	33	[Ar]$3d^{10}\ 4s^2\ 4p^3$
Ti	22	[Ar]$3d^2\ 4s^2$	Se	34	[Ar]$3d^{10}\ 4s^2\ 4p^4$
V	23	[Ar]$3d^3\ 4s^2$	Br	35	[Ar]$3d^{10}\ 4s^2\ 4p^5$
Cr	24	[Ar]$3d^5\ 4s^1$	Kr	36	[Ar]$3d^{10}\ 4s^2\ 4p^6$

1.6 Electronic configuration: box type representation

The electrons of an atom can also be represented by arrows drawn in boxes labelled for orbitals. A single box is used for an s orbital, a block of three boxes for a set of *p* orbitals and so on. Note some examples given below.

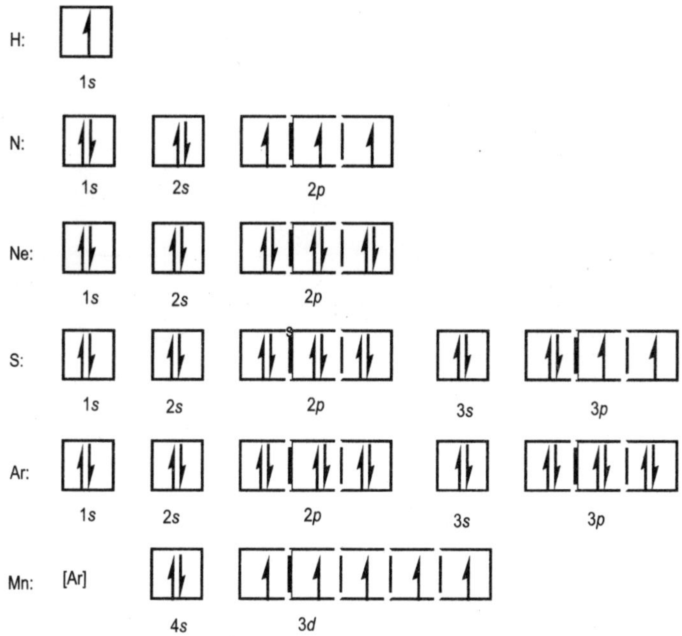

Figure 1.5: Electronic configuration: Box-type representation

Note: A filled atomic orbital contains two electrons with opposite spin and these are represented by two arrows, the heads of which point in opposite directions.

Practice question 1.3 Give the electronic configuration of the following elements together with their box-type representation.
(a) F ($Z=9$) (b) Al ($Z=13$) (c) K ($Z=19$) (d) V ($Z=23$)

1.7 The Periodic Table

There are 110 elements known to scientists. They are grouped and arranged in the form of a table called the **Periodic table.** The basis for the arrangement of elements in the modern periodic table is *atomic number* and *electronic configuration.* The elements are arranged in the order of increasing atomic number in horizontal rows in such a way that elements with similar outer shell electronic configuration fall in the same vertical column. The horizontal rows are called **Periods** and the vertical columns are called **Groups.**

The main features of the modern Periodic table are summarised here. There are

- Seven Periods labelled 1 to 7. These correspond to the seven energy levels in atoms.
- Eight main groups labelled Group I to VIII. Group I and II elements are called s-block elements and Group III – VIII, *p*-block elements.
- Four sets of transition elements called *d*-block elements which are placed in between Groups II and III in the Periodic table.
- Two sets of inner-transition elements called *f*-block elements, normally placed at the bottom of the periodic table (Table 1. 5).

Table 1.5: The main features of the Periodic Table

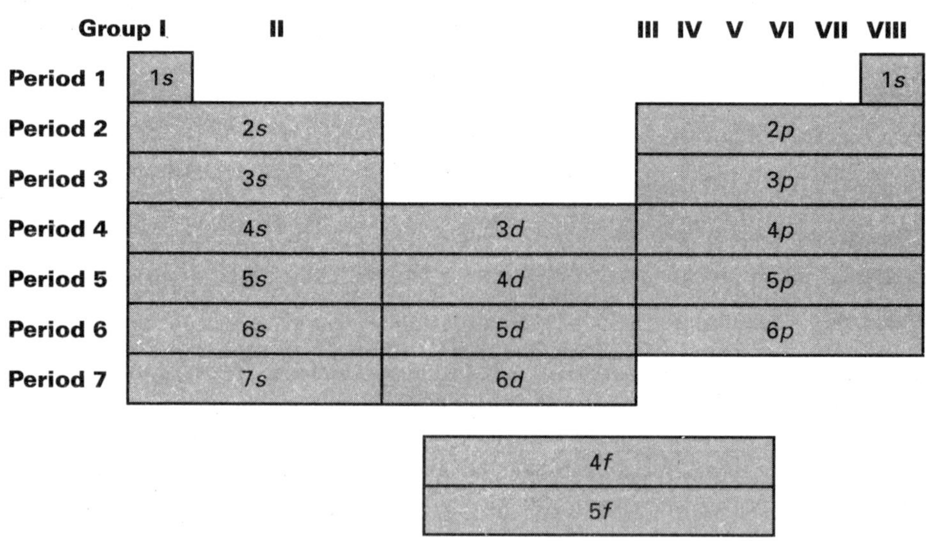

Period 1 consists of two elements; H ($Z=1, 1s^1$) and He ($Z=2, 1s^2$) and these correspond to the filling of the first energy level. Hydrogen is placed in Group I and helium, which has the first energy level complete, in Group VIII.

Periods **2** and **3** are short Periods and contain eight elements each; Li to Ne in Period 2, corresponding to the filling of the 2*s* and 2*p* orbitals and Na to Ar in Period 3 corresponding to the filling of 3*s* and 3*p* orbitals. There are two *s*-block and six *p*-block elements in each of these Periods (Table 1.6).

Table 1.6: Periods 2 and 3 elements

Period 2	3 Li [He]$2s^1$	4 Be [He]$2s^2$	5 B [He]$2s^22p^1$	6 C [He]$2s^22p^2$	7 N [He]$2s^22p^3$	8 O [He]$2s^22p^4$	9 F [He]$2s^22p^5$	10 Ne [He]$2s^22p^6$
Period 3	11 Na [Ne]$3s^1$	12 Mg [Ne]$3s^2$	13 Al [Ne]$3s^23p^1$	14 Si [Ne]$3s^23p^2$	15 P [Ne]$3s^23p^3$	16 S [Ne]$3s^23p^4$	17 Cl [Ne]$3s^23p^5$	18 Ar [Ne]$3s^23p^6$

Periods **4** and **5** are long Periods and consists of 18 elements each, eight corresponding to the filling of *s* and *p* orbitals (similar to Periods 2 and 3) and ten corresponding to the filling of *d* orbitals. Out of the 18 elements, two are *s*-block elements, six *p*-block and ten *d*-block elements.

Period **6** contains 32 elements; two *s*-block elements, six *p*-block elements, ten *d*-block elements and fourteen *f*-block elements.

Period **7** is similar to Period 6, but is incomplete.

Groups: Group I elements have one electron in the outermost energy level with an s^1 configuration, and the number of the period is the same as the number of the outermost energy level occupied. Similarly Group II elements have two electrons in the outer most energy level (s^2 configuration). Groups I and II are called *s*-block elements as *s* orbitals are being filled in these cases. The elements in Groups III to VIII are called *p*-block elements. Group III elements have three electrons (s^2p^1 configuration) in the outermost shell, Group IV have four (s^2p^2) and so on. Group VIII elements have eight electrons (s^2p^6).

Table 1.7

	Group I	**Group II**	**Group III —**	**Group VIII**
Period 2	Li [He]$2s^1$	Be [He]$2s^2$	B [He]$2s^22p^1$	Ne [He]$2s^22p^6$
Period 3	Na [Ne]$3s^1$	Mg [Ne]$3s^2$	Al [Ne]$3s^23p^1$	Ar [Ne]$3s^23p^6$
Period 4	K [Ar]$4s^1$	Ca [Ar]$4s^2$	Ga [Ar]$3d^{10}4s^24p^1$	Kr [Ar]$3d^{10}4s^24p^6$

1.8 Periodic properties

The chemical properties of elements depend largely on the number and type of electrons in the outermost energy level. Since all Group I elements contain an s electron in the outermost orbit, they have similar chemical properties; so do the members of other

Groups. We will study this resemblance in properties of elements of the same Group in Chapter 9.

It can be seen that the physical properties of elements vary gradually down a Group or along a Period. We will discuss some of these periodic properties here.

Atomic and ionic size (atomic and ionic radii)

The atomic size or atomic radius depends on the number of occupied energy levels in an atom and increases going down in a Group. This is understandable as more and more energy levels are added as we move down a Group. Consider Group I elements; Li ($1s^2 2s^1$) and Na ($1s^2 2s^2 2p^6 3s^1$). Li has electrons in two energy levels while sodium has them in three; sodium thus being larger. The atomic radius and ionic radius describe the distance from the nucleus to the outermost energy level.

Table 1.7: Atomic radii of Group I elements

Group I element	Atomic radius (nm)
Li	0.157
Na	0.191
K	0.235
Rb	0.250
Cs	0.272

In the same way, a sodium ion ($Na^+ = 1s^2 2s^2 2p^6$) is larger than a lithium ion ($Li^+ = 1s^2$).

Let us compare the atomic size of elements along a Period. Consider Period 2 elements; Li to Ne. Though the atomic number increases as we move along the Period, the atomic size (as well as the atomic radius) decreases (Table 1.8).

Table 1.8: Atomic radii of Period 2 elements

Period 2 element	Li	Be	B	C	N	O	F
Atomic radius (nm)	0.157	0.112	0.088	0.077	0.074	0.066	0.064

The gradual decrease in size is due to the increase in nuclear charge. As we move from left to right (of the period) the number of protons increases, therefore, the effective attraction of the nucleus for each electron increases. So the electrons are pulled closer to the nucleus and the size of the atom decreases.

Melting point

Melting points of elements, in general, increase along a period and decrease down a group. There are exceptions. The melting point of a substance depends on the type of intermolecular forces between the atoms (Chapter 3). Consider the melting points of Period 3 elements (Table 1.9). The melting point increases from sodium to silicon. Sodium, magnesium and aluminium are metals. The bonding between atoms in metals is described as *metallic bonding* which is the attraction between delocalised valence electrons of the metal atoms and positively charged atomic kernels. (See Section 3.7 on metallic bonding). The metallic bond strength increases from Na to Al as the number of valence electrons increases. Silicon is a non-metal, it has high melting point because in solid form, silicon has a giant covalent network lattice, similar to diamond (Section 3.8). Each silicon atom is bonded to four neighbouring silicon atoms by covalent bonds and this arrangement repeats in an infinite lattice. Melting requires breaking of these covalent bonds and thus a lot of energy is needed.

Phosphorus, sulphur and chlorine consist of covalent molecules; P_4, S_8 and Cl_2 respectively. The type bonding between the molecules in the solid state is *van der Waals forces*. Van der Waals forces are the attraction between neighbouring dipoles formed instantaneously. These forces of attraction between the molecules are weak; therefore lower melting points (See Section 3.8 on van der Waals forces). The strength of van der Waals forces depends on molecular weight. Since the molecular weight of S_8 is higher than that of P_4, sulphur has a higher melting point. The van der Waals forces between diatomic chlorine molecules is weaker. Argon is monatomic and the van der Waals forces are therefore weaker resulting in a lower melting point.

Period 3 element	Na	Mg	Al	Si	P	S	Cl	Ar
Melting point (K)	371	922	933	1683	317	386	172	84

Table 1.9: Melting points of Period 3 elements

Ionisation energy

Ionisation energy is the energy required to remove an electron from a gaseous atom. Ionisation energy increases along a period and decreases down a group. This is discussed in detail in Chapter 3 (Section 3.1).

Electron affinity

Electron affinity is the energy released when an electron is added to a gaseous atom of an element. Electron affinity also increases along a period and decreases down a group [Chapter 3 (Section 3.2)].

Electronegativity

Electronegativity is a measure of the power of attraction for a shared pair of electrons by two atoms in a covalent bond. Like ionisation energy and electron affinity, electronegativity also increases along a period and decreases down a group. Again this is discussed in detail in Chapter 3 (Section 3.6).

Practice question 1.4 Write the electronic configuration of

(a) Group I elements, Li ($Z = 3$), Na ($Z = 11$) and K ($Z = 19$), and

(b) Group VI elements, O ($Z = 8$), S ($Z = 16$), Se ($Z = 34$).

Tutorial: helping you learn

Progress questions

1.1 Name the three fundamental particles that constitute atoms of elements and give their relative masses and charges.

1.2 Write the number of protons, electrons and neutrons in an atom of each of the following.
(a) $^{40}_{20}Ca$ (b) $^{40}_{18}Ar$ (c) $^{31}_{15}P$ (d) $^{24}_{12}Mg$ (e) $^{30}_{15}P$ (f) $^{80}_{35}Br$

1.3 What are isotopes? Select the isotopes from the following list.
(a) $^{9}_{4}Be$ (b) $^{36}_{18}Ar$ (c) $^{30}_{15}P$ (d) $^{31}_{15}P$ (e) $^{40}_{18}Ar$ (f) $^{40}_{20}Ca$

1.4 Write the number of protons, electrons and neutrons in these isotopes of uranium.
(a) $^{235}U_{92}$ (b) $^{235}U_{92}$ (c) $^{238}U_{92}$

1.5 Write the electronic configuration of the following elements.
(a) P ($Z = 15$) (b) Si ($Z = 14$) (c) Mn ($Z = 25$)
(d) Ti ($Z = 22$) (e) Br ($Z = 35$)

1.6 Give the electronic configuration of the outermost orbit and the box-type representation of the following elements.
(a) O ($Z = 8$) (b) Si ($Z = 14$) (c) Mg ($Z = 12$)
(d) C ($Z = 6$) (e) P ($Z = 15$)

1.7 State, and illustrate where appropriate, in what respects, the following electrons in the orbitals of an atom may differ:
(a) 1*s* and 2*s* electrons (b) 2*s* and $2p_x$ electrons
(c) $2p_x$ and $2p_y$ electrons (d) two electrons in a $2p_x$ orbital

1.8 Arrange, giving reasons, the elements with atomic numbers 11, 15 and 19 in order of increasing atomic size.

2 Formulae, Equations and Stoichiometry

One-minute summary

We have seen in Chapter 1 that the masses of atoms and molecules are too small to express in grams. Scientists use the concept of relative atomic (or molecular) mass to express the mass of atoms and molecules as compared to the mass of a standard element. The isotope Carbon-12 (^{12}C) is used as the standard at present and is assigned a mass of 12 units in the atomic mass unit scale. A mole of a substance is a conveniently measurable quantity and in simple terms, it is the molecular mass of the substance expressed in grams. The mass spectrometer is an important tool in the determination of atomic and molecular masses. The mole concept, empirical and molecular formulae of compounds and worked out examples are given in this chapter.

This chapter explains

- relative atomic and molecular masses
- molar mass
- Avogadro's constant
- mass spectrometer
- empirical and molecular formulae
- calculations involving reacting masses
- calculations involving reacting volumes
- calculations involving molar concentrations of solutions

2.1 Formulae of chemical substances

Chemists represent elements by symbols and compounds by combinations of symbols, called formulae. The symbols of the elements are given in the Periodic table (page). The formula of a molecule of a substance (**molecular formula**) gives the number and type of atoms in one molecule of that substance. The formula of water is H_2O, i.e., a molecule of water consists of two atoms of hydrogen and one atom of oxygen. The formula of methane is CH_4 which means that a molecule of methane has four hydrogen atoms and one carbon atom.

The formula of an ionic compound (Section 3.3) describes the ratio of the different ions it contains. For example, the formula of sodium chloride is NaCl and sodium chloride consists of sodium ions (Na^+) and chloride ions (Cl^-) in the ratio 1:1. The formula of sodium sulphate is Na_2SO_4, which has two sodium ions (Na^+) for every sulphate ion (SO_4^{2-}).

2.2 Chemical equations

A **chemical equation** is used to represent a chemical reaction. It is a simple way of depicting the substances that take part in a reaction (**reactants**) and those that are formed (**products**). A chemical equation is written with the formulae of the reactants on the left hand side and the formulae of the products on the right hand side, separated by an arrow. For a reaction between hydrogen and oxygen to form water, the **word equation** is,

$$\text{Hydrogen} + \text{Oxyen} \longrightarrow \text{Water}$$

and the **chemical equation** is,

$$2H_2 + O_2 \longrightarrow 2H_2O$$

An equation must have the correct formulae of the reactants and products, and the number of atoms of all the elements in the two sides of the equation should be equal.

2.3 Relative atomic mass (A_r)

The **atomic mass unit scale** (noted by the symbol **amu** or **u** in recent times) is used to express the masses of atoms. Internationally, chemists use Carbon-12 (^{12}C) isotope as the standard, and ***one atomic mass unit, 1 u,*** *is defined as exactly one-twelfth the mass of an atom of* ^{12}C. Or, the atomic mass of an atom of ^{12}C is 12 u. (Note that, 1 u = 1.6605×10^{-24} g).

The **relative atomic mass, A_r,** of any other element is obtained in relation to the mass of a ^{12}C atom. For example, the relative atomic mass of hydrogen is 1.008 and that of chlorine is 35.453.

Note: Atomic mass is expressed in the atomic mass unit scale (u) and relative atomic mass, A_r has no units since it is a ratio. For example,
Atomic mass of ^{12}C = 12 u
Atomic mass of hydrogen = 1.008 u
Relative atomic mass of hydrogen = 1.008

Many naturally occurring elements consist of more than one isotope (Section 1.2). For example a naturally occurring sample of carbon contains 98.89% of ^{12}C of atomic mass 12.000 u and 1.11% of ^{13}C of atomic mass 13.003 u. The *average atomic mass* of a naturally occurring sample of carbon can be calculated as follows.

$$\text{Average atomic mass of carbon} = \frac{(98.89 \times 12.000\ \text{u}) + (1.11 \times 13.003\ \text{u})}{100}$$

Or,

$$= (0.9889 \times 12.00\ \text{u}) + (0.011 \times 13.0003\ \text{u})$$

$$= 12.01\ \text{u}$$

Example 2.1 Calculation of average atomic mass

Calculate the relative atomic mass, A_r of boron if boron consists of 19.78% of ^{10}B of isotopic mass 10.013 u and 80.22% of ^{11}B of isotopic mass 11.009 u.

Solution

$$A_r = (0.1978 \times 10.013) + (0.8022 \times 11.009) = 10.812$$

Relative atomic masses of some elements are given in Table 2.1. We shall make use of these atomic masses for the calculations.

Table 2.1: Relative atomic masses of some elements

Element	Symbol	A_r	Element	Symbol	A_r
Hydrogen	H	1.01	Aluminium	Al	26.98
Helium	He	4.00	Phosphorus	P	30.97
Boron	B	10.81	Sulphur	S	32.06
Carbon	C	12.01	Chlorine	Cl	35.45
Nitrogen	N	14.01	Argon	Ar	39.95
Oxygen	O	16.00	Potassium	K	39.10
Fluorine	F	19.00	Calcium	Ca	40.08
Neon	Ne	20.18	Manganese	Mn	54.94
Sodium	Na	22.99	Copper	Cu	63.55
Magnesium	Mg	24.31	Zinc	Zn	65.37

Practice question 2.1 Calculation of relative atomic mass

Calculate the relative atomic mass of chlorine if chlorine consists of two isotopes; 75.77% of ^{35}Cl of relative isotopic mass 34.9689 and 24.23% of ^{37}Cl of mass 36.9659.

2.4 Relative molecular mass (M_r)

Relative molecular mass, M_r, is obtained by adding the relative atomic masses of all the atoms in a molecule of that substance. For an ionic compound, the relative molecular mass is the same as the **relative formula mass** and is equal to the sum of the relative atomic masses of all atoms in the formula of the compound. Note the examples given below. (Relative atomic masses given in Table 2.1 are used in the calculations.)

$$M_r(H_2) = 2 \times 1.01 = 2.02$$

$$M_r(HCl) = 1.01 + 35.45 = 36.46$$

$$M_r(H_2O) = 2 \times 1.01 + 16.00 = 18.02$$

$$M_r(NaCl) = 22.99 + 35.45 = 58.44$$

$$M_r(CuSO_4 \cdot 5H_2O) = 63.55 + 32.06 + (4 \times 16.00) + (5 \times 18.02) = 249.71$$

Practice question 2.2 Calculation of relative molecular mass

Calculate the relative molecular mass of each of the following compounds.

(a) H_2SO_4 (b) C_3H_8 (c) $C_6H_{12}O_6$ (d) $MgCl_2 \cdot 6H_2O$

(Use Table 2.1 for relative atomic masses)

2.5 Molar mass and Avogadro's constant

1 mole (**1 mol**) of a substance is that quantity of it which contains as many particles (atoms, molecules or ions) as there are atoms in 12 g of ^{12}C. The number of atoms present in 12 g of ^{12}C is called **Avogadro's constant** and it is represented by the symbol ***L***. The value of Avogadro's constant is 6.022×10^{23}.

This means that 1 mole of ^{12}C contains 6.022×10^{23} carbon atoms and has a mass of 12 g. Similarly 1 mole of hydrogen atoms contains 6.022×10^{23} H atoms, 1 mole of hydrogen molecules contains 6.022×10^{23} H_2 molecules, and 1 mole of water contains 6.022×10^{23} H_2O molecules.

The **molar mass** of a substance is the mass of 1 mole of that substance and is expressed in g mol^{-1}

$$\text{Molar mass of H atoms} = 1.01 \text{ g mol}^{-1}$$
$$\text{Molar mass of } H_2 \text{ molecules} = 2.02 \text{ g mol}^{-1}$$
$$\text{Molar mass of } H_2O = 18.02 \text{ g mol}^{-1}$$

In other words, the molar mass of a substance is the molecular mass expressed in grams, and it contains 1 Avogadro's number of molecules.

The mass of a substance, the number of moles and the number of particles are thus related.

$$\text{Number of moles} = \frac{\text{Mass(g)}}{\text{Molar mass(g mol}^{-1})}$$

$$\text{Number of moles} = \frac{\text{Number of molecules}}{\text{Avogadro's constant}}$$

$$\text{Or, Number of moles} = \frac{\text{Number of molecules}}{6.022 \times 10^{23}}$$

2.6 Calculations involving moles and Avogadro's constant

Example 2.2 Mass-to-mole calculations

Calculate the number of moles of (1) H_2 in 5.0 g of hydrogen and (2) H_2O in 10.8 g of water. (A_r : H = 1, O = 16)

Solution

(1) Molar mass of H_2 = 2 g mol^{-1}

$$\text{Number of moles} = \frac{\text{Mass}}{\text{Molar mass}}$$

$$\text{Number of moles of } H_2 = \frac{5.0 \text{ g}}{2 \text{ gmol}^{-1}} = 2.5 \text{ mol}$$

(2) Molar mass of H_2O = 18 g mol^{-1}

$$\text{Number of moles of } H_2O = \frac{10.8 \text{ g}}{18 \text{ g mol}^{-1}} = 0.6 \text{ mol}$$

Note that, $\frac{\cancel{g}}{\cancel{g}\,\text{mol}^{-1}} = \frac{1}{\text{mol}^{-1}} = \text{mol}$

Example 2.3 Mole-to-mass calculations

Calculate the masses of the following substances, in grams.

(1) 1.2 mol of hydrogen chloride, HCl

(2) 0.05 mol of sodium hydroxide, NaOH

(A_r : H = 1, O = 16, Na = 23, Cl = 35.45)

Solution

Mass (g) = number of moles (mol) × molar mass (g mol^{-1})

(1) Molar mass of HCl = 1 + 35.45 = 36.45 g mol^{-1}

Mass of HCl in 1.2 mol = 1.2 mol × 36.45 g mol^{-1} = 43.74 g

(2) Molar mass of NaOH = 23 + 16 + 1 = 40 g mol^{-1}

Mass of NaOH in 0.05 mol = 0.05 mol × 40 g mol^{-1} = 2.0 g

Note that, $\cancel{\text{mol}} \times \text{g}\,\cancel{\text{mol}}^{-1} = \text{g}$

Example 2.4 Calculations involving Avogadro's constant

Calculate (1) the number of moles of chlorine atoms and (2) the number of chlorine atoms, in 3.2 g of chlorine.

(A_r : Cl = 35.45, Avogadro's constant, $L = 6.022 \times 10^{23}$)

Solution

Molar mass of Cl atom = 35.45 g mol^{-1}

(1) Number of moles of Cl atoms in 3.2 g of chlorine = $\frac{3.2\text{ g}}{35.45\text{ g mol}^{-1}} = 0.0903$ mol

(2) Numer of atoms in 1 mol = 6.022×10^{23}

Numer of Cl atoms in 0.0903 mol = $0.090.3 \times 6.022 \times 10^{23} = 5.438 \times 10^{22}$

Example 2.5 Calculations involving Avogadro's constant

Calculate (1) the number of moles of chlorine molecules and (2) the number of molecules, in 50.0 g of chlorine.

(A_r : Cl = 35.45, Avogadro's constant, $L = 6.022 \times 10^{23}$)

Solution

Molar mass of Cl_2 molecule = 35.45 × 2 = 70.90 g mol^{-1}

(1) Number of moles of Cl_2 molecules in 50.0 g of chlarine = $\frac{50.0\text{ g}}{70.90\text{ g mol}^{-1}} = 0.705$ mol

(2) Number of molecules in 1 mol = 6.022×10^{23}

Number of molecules in 0.705 mol = $0.705 \times 6.022 \times 10^{23} = 4.246 \times 10^{23}$

Practice question 2.3 Calculate (a) the number of moles and (b) the number of molecules, in 3.2 g of sulphur dioxide, SO_2 (A_r : O = 16, S = 32)

2.7 Mass spectrometry and the determination of atomic and molecular masses

The atomic mass, molecular mass, and relative abundance of various isotopes of an element can be determined with great accuracy using an instrument called a **mass spectrometer.**

In a mass spectrometer, the vaporised form of the substance is subjected to a beam of high energy electrons from an electron gun. Collision of a high energy electron with a molecule of the substance may knock out one or more electrons from it to form a positively charged particle. Removal of one electron from a molecule produces an ion with one positive charge. This ion is called a **molecular ion** and it has the same mass as that of the molecule. To knock out a second electron from a positive ion is difficult as a large amount of energy is required.

The positive ions, each of mass, ***m*** and charge, ***e*** are accelerated by passing through two oppositely charged electric plates and then through a magnetic field where the ions with the same ***m/e*** ratio take the same circular path while ions with different *m/e* ratios travel other paths. By adjusting the magnetic field strength, ions with a particular *m/e* ratio arrive at a detector where they produce an electric current. The intensity of the current depends on the number of ions. Thus the mass and abundance of one molecular ion is recorded on a chart. By varying the magnetic field strength, ions with other *m/e* ratios are detected and recorded.

The **mass spectrum** is a bar chart on which the relative intensity is plotted as the ordinate (y-axis) versus *m/e* as the abscissa (x-axis). The position of the peak denotes the mass. Look at the mass spectrum of chlorine (Figure 2.1). It can be seen that chlorine consists of two isotopes, ^{35}Cl and ^{37}Cl. The heights of the peaks depict their relative abundance.

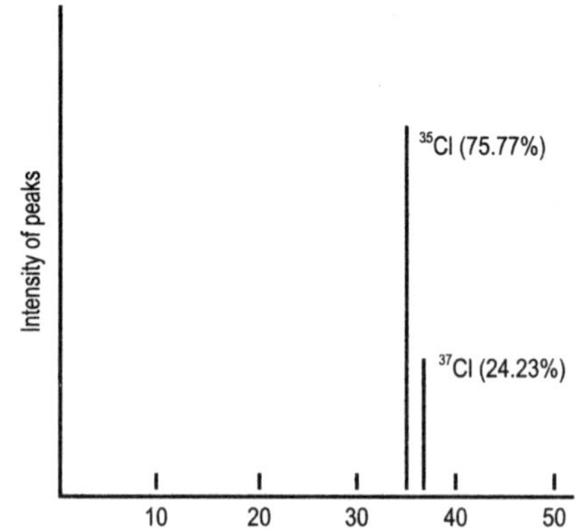

Figure 2.1: The mass spectrum of chlorine

2.8 Empirical and molecular formulae

The **molecular formula** of a compound is the formula which represents the type and number of all the atoms in a molecule of it. The molecular formula of water is H_2O, that of ethane is C_2H_6 and that of ethanoic acid (acetic acid) is $C_2H_4O_2$.

The **empirical formula** is the simplest formula which outlines the ratio of the number of atoms in a molecule of it. The empirical formula of water is the same as its molecular formula. The empirical formula of ethane is CH_3 since there are three H atoms for every C atom in the molecule, i.e., the ratio of C:H in ethane is 1:3. Similarly the empirical formula of ethanoic acid is CH_2O. The empirical formula of ethanoic acid shows that the ratio of C:H:O is 1:2:1. The molecular formula can be the empirical formula itself or a simple multiple (*n* times) of the empirical formula.

$$\text{Molecular formula} = (\text{Empirical formula})_n$$

where n is a simple whole number. Since a molecular formula contains n units of empirical formula, it follows that the molecular mass is n times the empirical formula mass. The value of n can thus be obtained from the equation,

$$n = \frac{\text{Molecular mass}}{\text{Empirical formula mass}}$$

The percentages of the elements in a compound can be determined experimentally. The empirical formula can be calculated from the percentage composition of the compound as shown in Example 2.6, and knowing the molecular mass, the molecular formula can be worked out.

Example 2.6 Empirical and molecular formulae

The percentages of carbon and hydrogen in ethanoic acid have been determined experimentally. Ethanoic acid contains 40.0% of C and 6.7% of H. The molecular mass of ethanoic acid is 60. Work out the empirical and molecular formulae of ethanoic acid. (A_r : H = 1, C = 12, O = 16)

Hint: If the percentages given do not add up to 100, this means that oxygen is present.

Solution

$$O = 100 - (40 + 6.7) = 53.3\%$$

Dividing the percentages of the elements by their respective atomic masses, we get the number of moles of atoms. The ratio of the number of moles of atoms is the same as the ratio of the number of atoms in the molecule. Dividing these numbers by the smallest number gives the simple whole number ratio of atoms in the molecule, as shown below.

Element	%	Atomic mass	%/at. mass	Divide by the smallest number
C	40.0	12	3.33	1
H	6.7	1	6.67	2
O	53.3	16	3.33	1

The empirical formula of ethanoic acid = CH_2O

Empirical formula mass of ethanoic acid = [12 + (2 × 1) + 16] = 30

$$n = \frac{\text{Molecular mass}}{\text{Empirical formula mass}} = \frac{60}{30} = 2$$

$$\text{Molecular formula} = (CH_2O)_2 = C_2H_4O_2$$

Practice question 2.4 A compound containing carbon, nitrogen and hydrogen has the percentage composition; C = 61.02%, N = 23.73%, H = 15.25%. The molecular mass of the compound is 59. Find the empirical and molecular formulae of the compound. (A_r : H = 1, C = 12, N = 14)

2.9 Reacting masses and calculations involving reacting masses

A chemical equation portrays the number of molecules that react and are formed. A *stoichiometric equation* (balanced chemical equation) can be used to calculate the masses of reacting substances or products.

Consider the reaction between hydrogen and oxygen to form water.

$$\underset{\text{2 molecules}}{2H_2(g)} + \underset{\text{1 molecule}}{O_2(g)} \longrightarrow \underset{\text{2 molecules}}{2H_2O(l)}$$

1 mole of any substance contains 1 Avogadro's number of molecules. It follows that 2 moles of hydrogen (4 g) react with 1 mole of oxygen (32 g) to form 2 moles of water (36 g).

$$\underset{\substack{\text{2 mol}\\ \text{4 g}}}{2H_2(g)} + \underset{\substack{\text{1 mol}\\ \text{32 g}}}{O_2(g)} \longrightarrow \underset{\substack{\text{2 mol}\\ \text{36 g}}}{2H_2O(l)}$$

Key points

1. Note that we get the mole ratio from the stoichiometric equation which allows us to calculate the reacting masses. Study the example given below.
2. At room temperature, hydrogen and oxygen are gases and water a liquid. It is common practice to include this information in the equation. Thus we use (g) for a gas, (l) for a liquid, (s) for a solid and (aq) for an aqueous solution.

Example 2.7 Calculations involving reacting masses

Magnesium burns in oxygen to form magnesium oxide. What mass of magnesium oxide is formed by burning 4.0 g of magnesium?
(A_r : O = 16, Mg = 24)

Solution

$$\text{Molar mass of MgO} = (24 + 16) = 40 \text{ g mol}^{-1}$$

$$\underset{2 \text{ mol}}{2Mg(s)} + O_2(g) \longrightarrow \underset{2 \text{ mol}}{2MgO(s)}$$

Mole ratio, Mg:MgO = 1:1

$$\text{Number of moles of Mg reacted} = \frac{\text{Mass}}{\text{Molar mass}} = \frac{4.0}{24} = 0.167 \text{ mol}$$

$$\text{Number of moles of MgO formed} = 0.167 \text{ mol}$$

$$\text{Mass} = \text{Number of moles} \times \text{molar mass}$$

$$\text{Mass of MgO formed} = 0.167 \text{ mol} \times 40 \text{ g mol}^{-1} = 6.68 \text{ g}$$

Example 2.8 Iron reacts with chlorine gas, under suitable conditions to form iron (III) chloride, ($FeCl_3$).

$$2Fe(s) + 3Cl_2(g) \longrightarrow 2FeCl_3(s)$$

Calculate the mass of iron (III) chloride that can be prepared starting with 2.5 g of iron. (A_r : Cl = 35.45, Fe = 55.85)

Solution

$$\underset{2 \text{ mol}}{2Fe(s)} + 3Cl_2(g) \longrightarrow \underset{2 \text{ mol}}{2FeCl_3(s)}$$

$$\text{Number of moles of iron} = \frac{2.5}{55.85} = 0.045 \text{ mol}$$

$$\text{Number of moles of } FeCl_3 = 0.045 \text{ mol}$$

$$\text{Molar mass of } FeCl_3 = 55.85 + (3 \times 35.45) = 162.2 \text{ g mol}^{-1}$$

$$\text{Mass of } FeCl_3 = 0.045 \text{ mol} \times 162.2 \text{ g mol}^{-1} = 7.30 \text{ g}$$

2.10 Molar volume

Molar volume is the volume occupied by 1 mole of any substance. The use of the molar volume is important for a gaseous substance.

The volume occupied by a gas, V dm^3, is related to the pressure exerted by the gas, P atmospheres (atm.), its number of moles, n, and the temperature, T K, by the equation,

$$PV = nRT$$

where R is the **ideal gas constant** and its value is 0.082 dm^3 atm K^{-1} mol^{-1}. The molar volume of a gas can be calculated at 1 atm. pressure and 20° C (293 K)

$$V = \frac{nRT}{P}$$
$$= \frac{1 \times 0.082 \times 293}{1}$$
$$= 24.03\ dm^3$$

Thus the value of the molar volume is 24.0 dm^3 (approximately) at 1 atm. pressure and 293 K and it is a constant for *any gaseous substance.*

2.11 Reacting volumes of gases

When gases react with each other, the volumes of the gases reacting and their products, if gaseous, bear a simple numerical ratio to one another, if all volumes are measured at the same temperature and pressure. This is known as ***Gay Lussac's law***.

Consider the reaction between hydrogen and chlorine to form hydrogen chloride. Experiments show that one volume of hydrogen combines with one volume of chlorine to form two volumes of hydrogen chloride, if all volumes are measured at the same temperature and pressure.

$$\underset{1\ vol}{H_2(g)} + \underset{1\ vol}{Cl_2(g)} \longrightarrow \underset{2\ vol}{2HCl(g)}$$

Note that 1 mole of hydrogen reacts with 1 mole of chlorine to form 2 moles of hydrogen chloride. 1 Mole of hydrogen and 1 mole of chlorine occupy 24.0 dm^3 each and 2 moles of hydrogen chloride 48.0 dm^3, all volumes measured at 1 atm pressure and 293 K. These are summarised as shown below.

$$\underset{\substack{1\ mol\\24.0\ dm^3\\1\ vol}}{H_2(g)} + \underset{\substack{1\ mol\\24.0\ dm^3\\1\ vol}}{Cl_2(g)} \longrightarrow \underset{\substack{2\ mol\\48.0\ dm^3\\2\ vol}}{2HCl(g)}$$

For a reaction involving gases, the volume ratio is the same as the mole ratio, provided all volumes are measured at the same temperature and pressure.

Example 2.9 Calculations involving volumes of gases
Calculate the volume of oxygen used and the volume of carbon dioxide formed when 2.0 dm^3 of methane is burnt in an excess of air, if all volumes are measured at the same temperature and pressure.

Solution

$$\underset{1\ vol}{CH_4(g)} + \underset{2\ vol}{2O_2(g)} \longrightarrow \underset{1\ vol}{CO_2(g)} + 2H_2O(l)$$

$$\text{Volume of oxygen reacted} = 4.0\ dm^3$$
$$\text{Volume of carbon dioxide formed} = 2.0\ dm^3$$

Example 2.10 Sodium carbonate reacts with hydrochloric acid according to the following equation,

$$Na_2CO_3(s) + 2HCl(aq) \longrightarrow 2NaCl(aq) + H_2O(l) + CO_2(g)$$

Calculate the mass of sodium chloride formed and the volume of carbon dioxide evolved at 1 atm. pressure and 293 K, when 5.3 g of sodium carbonate react with excess of hydrochloric acid. (A_r: C = 12, Na = 23, Cl = 35.45)

Solution

	$Na_2CO_3(s)$ + $2HCl(aq)$ $\longrightarrow$	$2NaCl(aq) + H_2O(l)$ +	$CO_2(g)$
	1 mol	2 mol	1 mol
Molar mass:	106 g mol^{-1}	58.45 g mol^{-1}	
Molar volume:			24.0 dm^3 at 1 atm. and 293 K

$$\text{Number of moles of } Na_2CO_3 = \frac{5.3 \text{ g}}{106 \text{ g mol}^{-1}} = 0.05 \text{ mol}$$

$$\text{Number of moles of NaCl} = 2 \times 0.05 = 0.1 \text{ mol}$$

$$\text{Mass of NaCl} = 0.1 \times 58.45 = 5.845 \text{ g}$$

$$\text{Number of moles of } CO_2 = 0.05 \text{ mol}$$

$$\text{Volume of } CO_2 = 0.05 \times 24.0 = 1.2 \text{ dm}^3 \text{ at 1 atm and 293 K}$$

2.12 Molarity of a solution and calculations involving molarities

Quantities of substances in solutions i.e., concentrations are expressed in **mole per dm^3 (mol dm^{-3})**. *A **molar solution** is a solution containing 1 mole of a substance in one dm^3 of the solution.* Normally we deal with aqueous solutions, that is a solute dissolved in water. The ***molarity*** of a solution is the number of moles of a substance present in a dm^3 of the solution. The unit of molarity is mole per dm^3.

$$\text{Molarity} = \frac{\text{Number of moles of solute}}{\text{Volume of solution } (dm^3)}$$

1 dm^3 of 1 molar (1 M) sodium hydroxide solution can be prepared by dissolving 40 g (1 mol) of sodium hydroxide in enough water to form 1 dm^3 solution. The molarity, or the molar concentration of this solution, is 1 mol dm^{-3}. If 4.0 g (0.1 mol) of sodium hydroxide is present in 1 dm^3 of solution, its molarity is 0.1 mol dm^{-3}.

Note: The volume of a solution prepared by dissolving 1 mole of a substance and 1 dm^3 of water need not be 1 dm^3. Molar concentration is the number of moles of the substance in 1 dm^3 of solution not in 1 dm^3 of water.

Example 2.11 Molarity or molar concentration
Calculate the molarity of a solution containing 14.0 g of sodium hydroxide in 250 cm^3 of the solution.

Solution

$$\text{Number of moles of NaOH} = \frac{\text{Mass}}{\text{Molar mass}}$$

$$= \frac{14.0}{40} = 0.35 \text{ mol}$$

$$\text{Volume of the solution} = \frac{250}{1000} = 0.25 \text{ dm}^3$$

$$\text{Molarity} = \frac{0.35\text{mol}}{0.25\text{dm}^3} = 1.4 \text{ mol dm}^{-3}$$

The relationship between molarity, the number of moles and the volume of the solution can be used to calculate anyone of the three variables, provided the other two are given. See the examples 2.12–2.14.

Example 2.12 Calculate the number of moles of HCl present in 50 cm^3 of 2.0 mol dm^{-3} solution of hydrochloric acid.

Solution

$$\text{Volume of solution} = \frac{50}{100} = 0.05 \text{ dm}^3$$

$$\text{Number of moles} = \text{Molarity} \times \text{Volume}$$

$$= 2.0 \text{ mol dm}^{-3} \times 0.05 \text{ dm}^3 = 0.1 \text{ mol}$$

Example 2.13 What volume of 0.4 mol dm^{-3} solution that can be prepared using 10.0 g of potassium hydroxide. (A_r: K = 39, O = 16, H = 1)

Solution

$$\text{Molar mass of KOH} = 56 \text{ g mol}^{-1}$$

$$\text{Number of moles} = \frac{10.0 \text{ g}}{56 \text{ g mol}^{-1}} = 0.18 \text{ mol}$$

$$\text{Volume of solution (dm}^3) = \frac{\text{number of moles}}{\text{Molarity}}$$

$$= \frac{0.18 \text{ mol}}{0.4 \text{ mol dm}^{-3}} = 0.45 \text{ dm}^3$$

Example 2.14 Calculate the mass of sodium carbonate required to prepare 100 cm^3 of 0.2 mol dm^{-3} solution. (M_r: Na_2CO_3 = 106)

Solution

Molarity of the solution = 0.2 mol dm^{-3}

Volume of solution required = 0.1 dm^3

Number of moles required = 0.2 × 0.1 = 0.02 mol

Molar mass of Na_2CO_3 = 106 g mol^{-1}

Mass of Na_2CO_3 = 0.02 mol × 106 g mol^{-1} = 2.12 g

Tutorial: helping you learn

Progress questions

2.1 Calculate the average atomic mass of a sample of uranium, if it consists of 0.7% of ^{235}U of relative atomic mass 235.044 and 99.3 % of ^{238}U of relative atomic mass 238.051.

2.2 Calculate the number of moles in each of the following samples.
(a) 8.0 g He (b) 0.05 g $CaCO_3$
(c) 1.0×10^{25} atoms of iron (d) 9.033×10^{21} molecules of CO_2
(e) 5.6 dm^3 of nitrogen (II) oxide, NO, at 1 atm. and 293 K
(A_r : He = 4, C = 12, O = 16, Ca = 40, Avogadro's constant, $L = 6.022 \times 10^{23}$, Molar volume = 24.0 dm^3 at 1 atm. and 293 K)

2.3 Arrange the following substances in order of increasing numbers of molecules.
(a) 5.5 g of CO_2 (b) 2.5 g of H_2 (c) 10.0 g of N_2
(d) 50.0 g of H_2SO_4 (e) 80.0 g of C_2H_6 (f) 0.25 g of He
(A_r : H = 1, He = 4, C = 12, N = 14, O = 16, S = 32)

Hint: In each of the above, the order of number of moles and the number of molecules are the same.

2.4 Arrange the following in order of increasing numbers of molecules.
(a) 3.0×10^{20} H_2O molecules (b) 5.0 dm^3 of CO_2 at 1 atm and 293 K
(c) 20.0 g of NH_3 (d) 9.6 dm^3 of SO_2 at 1 atm and 293 K
(A_r : H = 1, N = 14, Avogadro's constant, $L = 6.022 \times 10^{23}$, Molar volume = 24.0 dm^3 at 1 atm and 293 K)

2.5 A hydrocarbon of relative molecular mass of 56 contains 85.71% carbon and 14.29% of hydrogen. Find the empirical and molecular formula of the compound. (A_r : H = 1, C = 12)

2.6 Sulphur is converted to sulphuric acid, industrially, by the following reactions.

$$S(s) + O_2(g) \longrightarrow SO_2(g)$$
$$2SO_2(g) + O_2(g) \longrightarrow 2SO_3(g)$$
$$SO_3(g) + H_2O(l) \longrightarrow H_2SO_4(l)$$

Calculate the theoretical mass of sulphuric acid that can be prepared starting from 160.0 g of sulphur. (A_r : H = 1, O = 16, S = 32)

Hint: The mole ratio of S:H_2SO_4 = 1:1

2.7 Calculate the concentration of each of the following in mol dm^{-3}.
(a) 2.0 g of NaCl in 100 cm^3 solution
(b) 5.0 mole of H_2SO_4 in 8 dm^3 solution
(c) 3.16 g of $KMnO_4$ in 200 cm^3 solution

(d) 8.4 g of $NaHCO_3$ in 2.5 dm^3 solution
(A_r : H = 1, C = 12, O = 16, Na = 23, Cl = 35.45, K = 39, Mn = 55)

2.8 100 cm^3 of 2.0 mol dm^{-3} hydrochloric acid is treated with excess magnesium metal. Calculate
(a) the number of moles of HCl taken
(b) the number of moles of magnesium reacted
(c) the mass of magnesium
(A_r : Mg = 24)

2.9 For the reaction between nitrogen and hydrogen to form ammonia,

$$N_2(g) + 3H_2(g) \longrightarrow 2NH_3(g)$$

calculate

(a) the volume of hydrogen that would react with 3 dm^3 of nitrogen and

(b) the volume of ammonia that would be formed, if all volumes are measured at the same temperature and pressure

3 Chemical Bonding and States of Matter

One-minute summary

The three main types of bonds in compounds are electrovalent/ionic, covalent and co-ordinate. Ionic bonding is the attraction between positive and negative ions formed by the reaction of metals with non-metals. The ease with which an ionic compound is formed depends on the ionisation energy of the metal and the electron affinity of the non-metal. Covalent bonds are formed between atoms by the sharing of a pair of electrons and co-ordinate bonds by unequal sharing. Most compounds cannot be classified as pure ionic or pure covalent compounds. There is some ionic and some covalent character in most compounds, the extent of each depending on the electronegativities of the atoms. There are other types of bonding; metallic bonding in metals, hydrogen bonding between molecules containing hydrogen bonded to a highly electronegative element, dipole-dipole attraction between polar molecules and van der Waals forces in all substances. Some of these intermolecular forces are responsible for holding molecules in a liquid or solid state.

Topics discussed in this chapter include:

- ionisation energy
- electron affinity
- ionic bonding
- covalent bonding
- electronegativity
- co-ordinate bonding
- metallic bonding
- hydrogen bonding
- van der Waals attraction
- dipole-dipole attraction

3.1 Ionisation energy

Energy is needed to remove one or more electrons from an atom. The **first ionisation energy** is defined as the energy required to remove the most loosely held electron from each of 1 mole of gaseous atoms in its ground state (the state of lowest energy). Ionisation energy is positive (endothermic) and is expressed in kJ mol^{-1} (Section 4.1).

For example, the first ionisation energy of sodium is +496.0 kJ mol^{-1}. This means that 496.0 kJ of energy is required to convert 1 mole of gaseous Na atoms into gaseous Na^+ ions. Note that the most loosely held electron is removed first and in the case of Na ($1s^2 2s^2 2p^6 3s^1$), the electron from the third energy level.

First ionisation energy of sodium,

$$Na(g) \longrightarrow Na^+(g) + e^-, \quad \Delta H_{I.E.} = +496.0 \text{ kJ mol}^{-1}$$

The energy required to remove a second electron from the positive ion created above is the **second ionisation energy**, the second ionisation energy always being higher than the first. The attraction between the nucleus and the electrons is stronger in a positive ion than in a neutral atom. So it is more difficult to remove an electron from a positive ion and more energy is required for the second ionisation.

Second ionisation energy of sodium,

$$Na^{+}(g) \longrightarrow Na^{2+}(g) + e^{-}, \quad \Delta H_{I.E.} = +4563.0 \text{ kJ mol}^{-1}$$

Ionisation energies of elements depend on a number of factors.

1. **Nuclear charge:** As nuclear charge (atomic number) increases, ionization energy increases, since there is greater attraction by the more positively charged nucleus for the electron to be removed.
2. **Atomic size:** As atomic size increases, ionisation energy decreases. This is because the electron to be removed is further from the nucleus and so held by a weaker attracting force.
3. **Screening effect:** The screening effect or the repulsion of the inner shell electrons for the outer shell electrons lowers ionisation energy.
4. **The type of the electron to be removed:** It is more difficult to remove an *s* electron than to remove a *p* electron from the same energy level (Section 1.5). The ease of removing an electron from any one energy level decreases in the order, s > p > d > f. Again it is easier to remove a single unpaired electron from an orbit than to remove one from a pair.

Ionisation energy as a periodic property

Ionisation energy is a periodic property. *Ionisation energy increases along a Period (with a few exceptions) and decreases down a Group.* The increase in ionisation energy along a Period is due to the increase in the effective nuclear charge which in turn depends on the increase in nuclear charge and decrease in atomic size (Section 1.8). The ionisation energies of Period 2 elements are shown for comparison (Table 3.1).

Table 3.1: Ionisation energies of Period 2 elements

Period 2 element	Li	Be	B	C	N	O	F	Ne
Nuclear charge	3	4	5	6	7	8	9	10
Atomic radius (nm)	0.191	0.125	0.090	0.077	0.075	0.073	0.071	0.065
1st I.E. (kJ mol^{-1})	520	899	801	1086	1400	1314	1680	2080

There are two exceptions to the increasing of ionisation energy along Period 2.

1. The ionisation energy of boron, B, ($1s^2 2s^2 2p^1$) is lower than that of beryllium, Be, ($1s^2 2s^2$) although B has a higher nuclear charge and smaller size than Be. Boron has a single electron in the outermost shell ($2p^1$) which is shielded by the inner $2s^2$ electrons. It is also easier to remove a single *p* electron (in B) than a paired s^2 electron (in Be).

2. The ionisation energy of oxygen ($1s^2 2s^2 2p^4$) is lower than that of nitrogen ($1s^2 2s^2 2p^3$). In nitrogen the three *p* electrons are in separate *p* orbitals ($p_x^1 p_y^1 p_z^1$). This is a more stable arrangement than the arrangement of four electrons in the three *p* orbitals of the oxygen atom ($p_x^2 p_y^1 p_z^1$) where a *p* orbital has a pair of electrons and there is more repulsion between electrons.

Similar trends and exceptions can be found in the other Periods.

Ionisation energy decreases down a Group. Take the example of Group 1 elements (Table 3.2).

Group I element	1st I.E. (kJ mol^{-1})
Li	520
Na	496
K	419
Rb	403
Cs	376

Table 3.2: Ionisation energies of Group I elements

Though nuclear charge increases down the Group, atomic size increases as more and more orbits are added, so the electron to be removed is farther from the nucleus and is less tightly held. There is also the increasing shielding effect between the nucleus and the outer electron by the complete inner shell electrons as we progress down the group. All the Group I elements have the same outer shell configuration, s^1. As the atomic size and shielding effect increases, the net attraction between the nucleus and the electron to be removed decreases and so ionisation energy decreases down a Group.

Example 3.1

(a) The 1st ionisation energy of calcium is higher than that of potassium.

(b) The 2nd ionisation energy of calcium is much lower than that of potassium.

Give explanations for these observations.

	Atomic number	1st I.E (kJ mol^{-1})	2nd I.E. (kJ mol^{-1})
Potassiun	19	419	3051
Calcium	20	590	1145

Table 3.3

Solution

(a) K: $1s^2\ 2s^2 2p^6\ 3s^2 3p^6\ 4s^1$ Ca: $1s^2\ 2s^2 2p^6\ 3s^2 3p^6\ 4s^2$

The effective nuclear charge of Ca is higher than that of K and so the electron to be removed is held more strongly in Ca. Also, it is more difficult to remove a paired *s* electron ($4s^2$) from Ca than to remove a single s electron ($4s^1$) from K.

(b) K^+: $1s^2\ 2s^22p^6\ 3s^23p^6$ Ca^+: $1s^2\ 2s^22p^6\ 3s^23p^6\ 4s^1$
The 2nd ionisation energy of Ca is the energy needed to remove the single $4s^1$ electron in Ca^+ which is well shielded by the inner shell electrons. It is easier to remove this electron than one from the complete inner shell ($3p^6$) in K^+.

Practice question 3.1 The 1st ionisation energy of B ($Z = 5$) is 736 kJ mol^{-1} and that of Al ($Z = 13$) is 577 kJ mol^{-1}. Give an explanation for the difference in the two values.

3.2 Electron affinity

The **first electron affinity** is the energy released when an electron is added to each of 1 mole of gaseous atoms to form gaseous ions. The unit for electron affinity is kJ mol^{-1}. The electron affinity of chlorine is given by the equation,

$$Cl(g) + e^- \longrightarrow Cl^-(g), \quad \Delta H_{E.A.} = -348.0 \text{ kJ mol}^{-1}.$$

This means that 348.0 kJ of energy is evolved when 1 mole of gaseous chlorine atoms are converted into gaseous chloride ions. *Electron affinity increases along a Period as the nuclear charge increases and atomic radius decreases.* Elements in Groups VI and VII have a greater tendency to accept electrons and become stable, and the value for the first electron affinity is high for these elements (Table 3.4). Note that the higher the negative value of the electron affinity, the greater the tendency for the element to take an electron.

Table 3.4: The electron affinities of some elements

	Group V	Group VI	Group VII
Element	N	O	F
E.A. (kJ mol^{-1})	+7	–140	–328

Electron affinity decreases down a Group as the atomic size increases. Compare the electron affinities of Group VII elements, chlorine, bromine and iodine (Table 3.5).

Table 3.5

Group VII element	E.A. (kJ mol^{-1})
Cl	–348
Br	–325
I	–295

When a halogen atom accepts an electron, it gets the stable electron configuration of an inert gas with filled *s* and *p* orbitals. For example, chlorine ($1s^2 2s^2 2p^6 3s^2 3p^5$) accepts an electron and forms Cl^- ($1s^2 2s^2 2p^6 3s^2 3p^6$) with the electron configuration of Ar. Inert gases have no tendency to accept electrons.

Example 3.2

The first electron affinity of oxygen is –140 kJ mol^{-1} and its second electronaffinity is +840 kJ mol^{-1}. (1) Give an explanation for the high positive value of its 2nd electron affinity.

(2) What is the total energy needed to convert 1 mole of O (g) to O^{2-} (g)?

Solution

(1) Energy is released (E.A. negative) when an electron is added to an oxygen atom to form an O^- ion. Energy is needed to force a second electron to a negative O^- ion because of the natural repulsion from the negative ion.

(2) The total energy required to convert 1 mole of O (g) to O^{2-} (g) = (–140 + 840) = +700 kJ mol^{-1}.

3.3 Ionic bonding and properties of ionic compounds

Most compounds formed between metals and non-metals are ionic compounds. Metals generally have low ionisation energies, and non-metals, high electron affinities. Metal atoms donate electron/s and form positive ions (cations), and non-metal atoms accept electron/s and form negative ions (anions). The electrostatic force of attraction between positive and negative ions is called **ionic bonding.**

Let us study the formation of an ionic bond taking sodium fluoride as an example. Sodium combines with fluorine to form sodium fluoride.

$$2\text{Na(s)} + \text{F}_2\text{(g)} \longrightarrow 2\text{NaF(s)}$$

What fundamental changes take place during this reaction? The sodium atom ($1s^22s^22p^63s^1$) has one electron in its outermost orbit and the fluorine atom has seven ($1s^22s^22p^5$). When sodium reacts with fluorine, the sodium atom donates its valence electron ($3s^1$) and becomes a positive ion (Na^+) which has a stable outer shell configuration of s^2p^6. The fluorine atom takes this electron to form a negative ion (F^-) which again becomes stable with an outer shell configuration of s^2p^6. The oppositely charged sodium and the fluoride ions attract each other in sodium fluoride.

$$\begin{array}{ccc} \text{Na} & \longrightarrow & \text{Na}^+ + \text{e}^- \\ 1s^2 2s^2 2p^6 3s^1 & & 1s^2 2s^2 2p^6 \\ \text{F} + \text{e}^- & \longrightarrow & \text{F}^- \\ 1s^2 2s^2 2p^5 & & 1s^2 2s^2 2p^6 \\ \text{Na}^+ + \text{F}^- & \longrightarrow & \text{Na}^+\text{F}^- \end{array}$$

This force of attraction between oppositely charged ions is called ionic bonding.

When sodium combines with oxygen to form sodium oxide, two sodium atoms give one electron each to form two sodium ions and an oxygen atom ($1s^22s^22p^4$) takes these two electrons to form an oxide ion, O^{2-} ($1s^22s^22p^6$).

$$\begin{array}{ccc} 2\text{Na} & \longrightarrow & 2\text{Na}^+ + 2\text{e}^- \\ \text{O} + 2\text{e}^- & \longrightarrow & \text{O}^{2-} \\ 2\text{Na}^+ + \text{O}^{2-} & \longrightarrow & \text{Na}_2\text{O} \end{array}$$

Note:

1. The formula of sodium oxide is Na_2O as two Na atoms donate an electron each and become Na^+ ions and one O atom accepts two electrons to form O^{2-}. We can write the formulae of ionic compounds, knowing the number of valence electrons in atoms and also the fact that the total charge of the ions in the formula should be zero.
2. A **Lewis electron dot structure** depicts the outermost shell electrons (valence electrons) as dots or crosses. The valence electrons are responsible for bond formation and the complete inner shells remain without any changes. The Lewis structures of some atoms are given below.

H· Li· Na· ·N̈· :Ö· :F̤̈· Mg: Äl·

Example 3.3 Use dots and/or crosses (Lewis structure) to portray the bond formation in
(1) magnesium chloride and (2) aluminium oxide.

Solution

(1)

$$Mg{:} \longrightarrow Mg^{2+} + 2e^-$$

$$2\,{:}\ddot{\underset{..}{Cl}}\cdot + 2e^- \longrightarrow 2\,{:}\ddot{\underset{..}{Cl}}{:}^-$$

$$Mg^{2+} + 2Cl^- \longrightarrow MgCl_2$$

(2)

$$2\dot{Al}{:} \longrightarrow 2Al^{3+} + 6e^-$$

$$3\cdot\ddot{\underset{..}{O}}\cdot + 6e^- \longrightarrow 3\,{:}\ddot{\underset{..}{O}}{:}^{2-}$$

$$2Al^{3+} + 3O^{2-} \longrightarrow Al_2O_3$$

Properties of ionic compounds

1. Ionic compounds have crystalline lattices where ions are arranged usually in a face centred cubic (FCC) lattice or a body centred cubic (BCC) lattice (Section 3.11).
2. Ionic compounds have high melting and boiling points. Ionic bonds are strong and a large amount of energy is needed to break the strong bonds between ions so that the ions move out from lattice points. There is free movement of particles in the liquid or gaseous state; that is when melted or boiled.
3. Ionic solids conduct electricity in the molten form or when dissolved in water. In both cases, there are ions moving which is necessary for the conduction of electricity.
4. Ionic solids are generally soluble in water. Water molecules break the ions loose from the solid lattice and bond with them. Each cation bonds with four or six water molecules through the negative oxygen end and each anion through the positive hydrogen end of the water molecules.

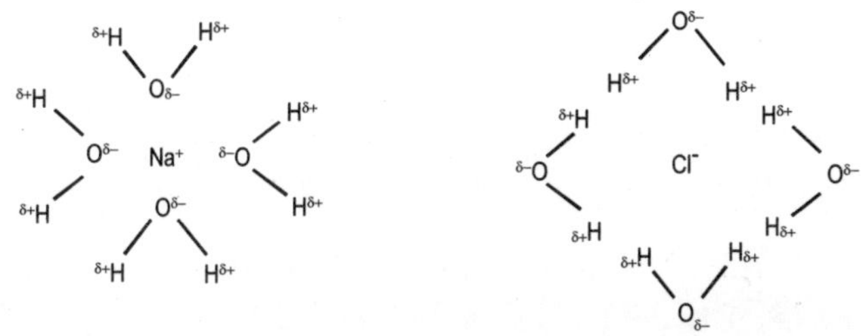

Figure 3.1: Water molecules surrounding Na^+ and Cl^- ions in sodium chloride solution

Example 3.4 Arrange the substances in order of increasing melting points, and relate your sequence to the forces holding units together in the solid state: $CaCl_2$ (s), KCl (s) and HCl (s).

Solution

$$\underrightarrow{\text{HCl (s)} < \text{KCl (s)} < \text{CaCl}_2\text{(s)}}_{\text{increasing melting point}}$$

KCl and $CaCl_2$ are ionic solids and the electrostatic forces of attraction between positive and negative ions in these solids are stronger than the hydrogen bonding between molecules of HCl. The ionic bonding in $CaCl_2$ is stronger than that in KCl since the attraction between doubly positively charged Ca^{2+} and Cl^- ions is stronger than that between singly positive K^+ and Cl^- ions.

3.4 Covalent bonding

The bonding between atoms in molecules like H_2, Cl_2, HCl and CO_2 is covalent. Let us study covalent bond formation taking H_2 as an example. A hydrogen atom has one electron ($1s^1$) in its outer orbit. When two hydrogen atoms come close enough, the nucleus of one atom attracts the electron of the other atom and *vice versa*. The two electrons become 'a shared pair' and are attracted by both nuclei. The 'shared electron pair' becomes a common property of both atoms, holding the two nuclei together. This type of bonding between two atoms by a shared electron pair is called **covalent bonding**. Each atom of hydrogen in the molecule attains the stable electron configuration of helium with two shared electrons.

$$\text{H}\cdot + \text{H}\cdot \longrightarrow \text{H:H (H—H)}$$

A Cl_2 molecule is formed when two chlorine atoms each with seven electrons in the outer-most orbit ($1s^2 2s^2 2p^6 3s^2 3p^5$) share two electrons to form a 'shared pair' which holds the atoms together.

$$:\ddot{\underset{..}{\text{Cl}}}\cdot + :\ddot{\underset{..}{\text{Cl}}}\cdot \longrightarrow :\ddot{\underset{..}{\text{Cl}}}:\ddot{\underset{..}{\text{Cl}}}:$$
$$\text{Cl—Cl}$$

Notes

1. A covalent bond is represented by a hyphen, i.e., a line drawn between two atoms to represent a shared pair of electrons.
2. Only a single (unpaired) electron in an orbit can take part in covalent bond formation.
3. Each atom involved in the bond normally gets the stable outer shell configuration of an inert gas (s^2p^6). Hydrogen attains an s^2 configuration.
4. Normally, non-metals which have four or more electrons in their outer shells undergo covalent bond formation.

Example 3.5

Use dots and/or crosses to depict the structures of the following molecules

(1) HCl (2) NH_3 (3) CO_2 (4) PCl_5

Solution

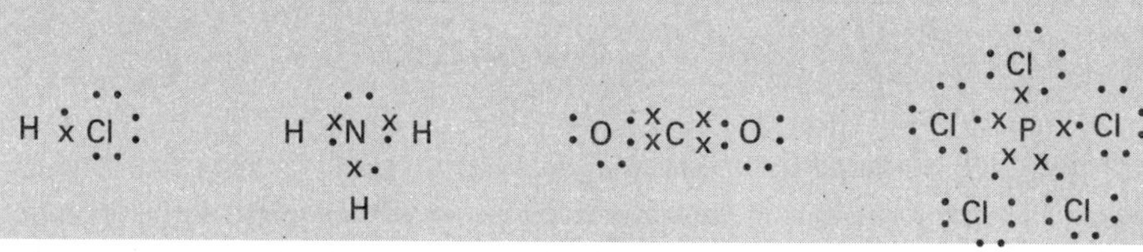

Example 3.6

Write the structural formulae (displayed formulae) of each of the compounds in Example 3.5.

Solution

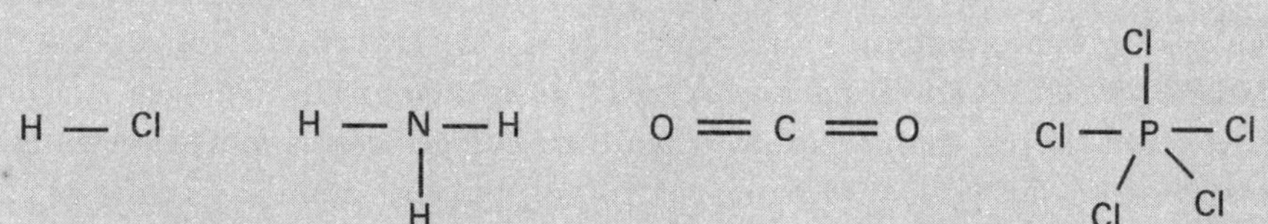

Overlapping of atomic orbitals

Covalent bond formation can be explained in terms of 'overlapping of atomic orbitals'. Let us look at the formation of a H_2 molecule. Two *s* orbitals of two hydrogen atoms overlap or merge to form a molecular orbital and the two electrons from the two atomic orbitals remain in this molecular orbital. This is shown diagrammatically in Figure 3.2.

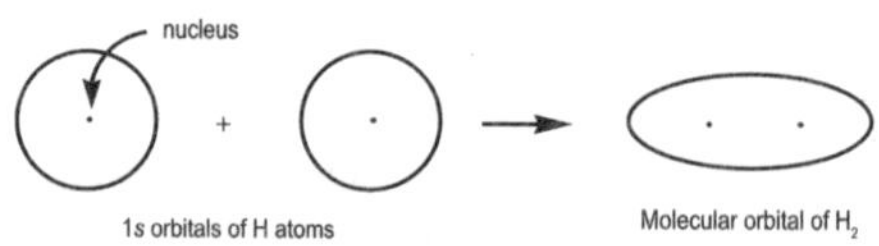

Figure 3.2: Overlapping of atomic orbitals of hydrogen

A molecular orbital, like an atomic orbital, can take a maximum of two electrons. The bond formed is called a **sigma(σ) covalent bond**. The extent of overlapping of two *s* orbitals is high, thus the σ covalent bond is strong. In an HCl molecule, the *s* orbital of hydrogen and the half-filled *p* orbital of chlorine overlap (Figure 3.3), and in a Cl_2 molecule, two half-filled *p* orbitals of two chlorine atoms (Figure 3.4). In both these cases σ covalent bonds are formed. We will study covalent bond formation in more detail in Chapter 8.

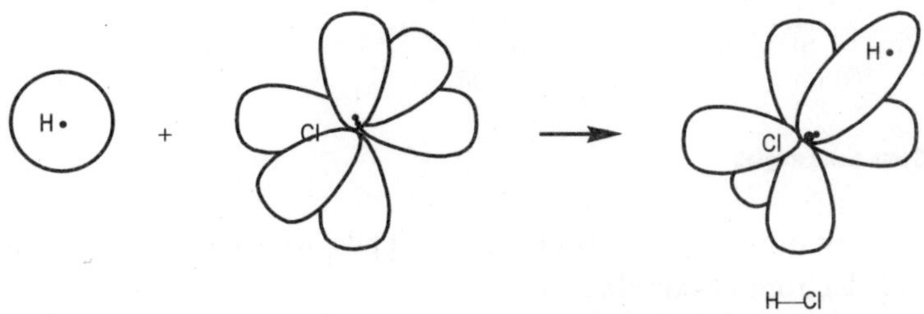

Figure 3.3: Formation of HCl

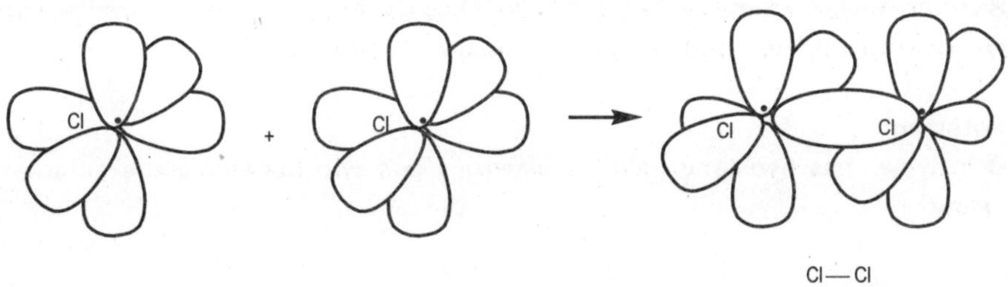

Figure 3.4: Formation of Cl_2

Properties of covalent compounds

1. Covalent compounds usually have low melting and boiling points. Although covalent bonds are strong, the bonds *between the molecules* in covalent compounds are weak van der Waals forces or dipole-dipole attraction. Melting and boiling points depend on the bonds between molecules and not on the bonds within the molecule. Thus many covalent compounds are liquids or gases.
2. They are generally insoluble in water but soluble in solvents like diethyl ether and trichloromethane. The solubility in such solvents is due to the ability of covalent compounds to form bonds of similar strength as present between molecules of the solvent.
3. They are not electrical conductors as there are no ions.

Bond Length is the distance between the centres of the two atomic nuclei which form the bond. The bond length of the H-H molecule is 0.074 nm, that of H-Cl is 0.127 nm and that of H-O in H_2O is 0.096 nm.

Bond angle is the angle between the two lines formed by joining the centres of the atoms concerned. For example, the bond angle in H_2O is 104.5°, that in NH_3 is 107° and the bond angle in CO_2 is 180° (Figure 3.5).

Bonding pair and lone pair of electrons

The shared electron pair in a covalent bond is the *bonding pair* while the electron pair/s

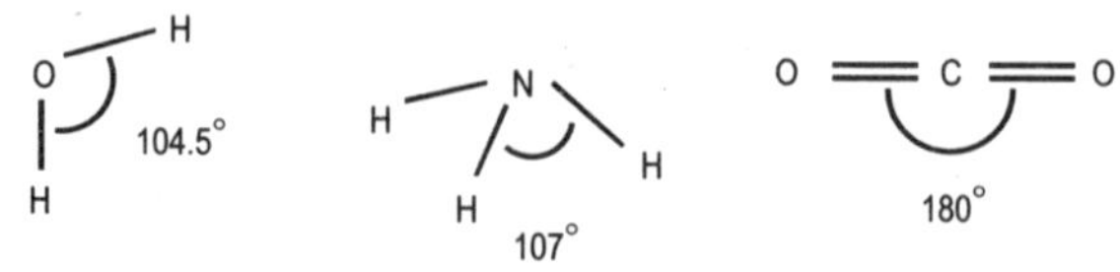

Figure 3. 5: Some bond angles

in the outermost orbit of an atom which do not take part in the bonding are called the *lone pairs* of electrons (Example 3.7).

Example 3.7

Draw the bonding pair and the lone pairs of electrons in a molecule of water.

Hint: An oxygen atom ($1s^2\ 2s^2 2p_x^{\ 2} 2p_y^{\ 1} 2p_z^{\ 1}$) has six electrons in its outermost orbit; two pairs of electrons and two single (unpaired) electrons.

Solution

O in water has two lone pairs of electrons and two bonding pairs as shown below.

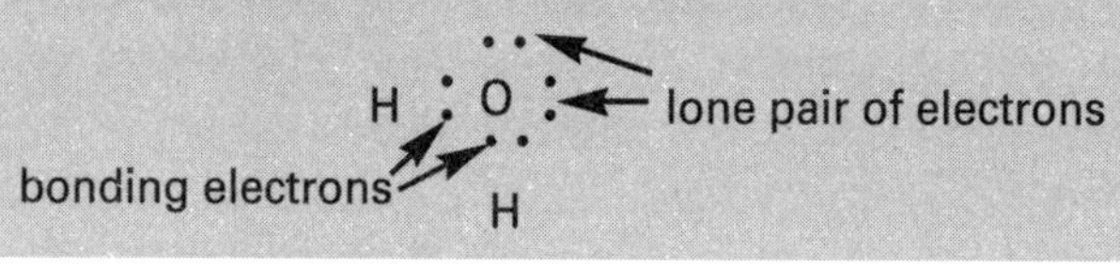

3.5 Co-ordinate bonding

When two atoms share two electrons and if these two electrons which form the bond come from one of the two atoms and share with a vacant orbital of the other atom, the bond formed is called a **dative covalent** bond or a **co-ordinate** bond. NH_4Cl and $F_3B.NH_3$ are examples of compounds containing co-ordinate bonds. When ammonia combines with hydrogen chloride, ammonium chloride ($NH_4^+Cl^-$) is formed. Here the nitrogen forms a co-ordinate bond with H^+ using its lone pair of electrons, to form NH_4^+ (Figure 3.6). A co-ordinate bond is represented by an arrow, the head of the arrow pointing to the acceptor.

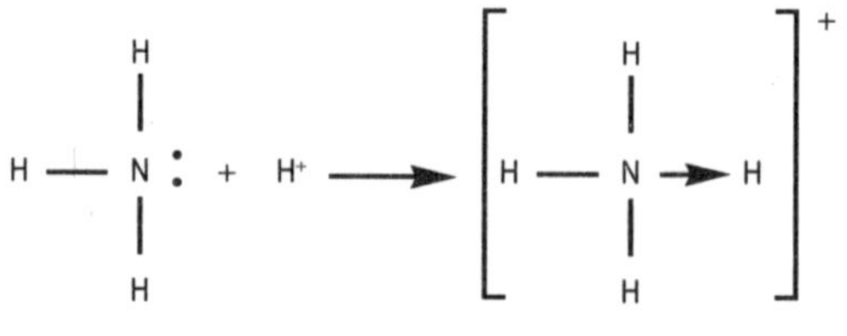

Figure 3.6: Co-ordinate bond formation in NH_4^+ ion

When boron trifluoride reacts with ammonia to form $F_3B \cdot NH_3$, a co-ordinate bond is formed between boron and nitrogen using the lone pair of electrons on nitrogen (Figure 3.7).

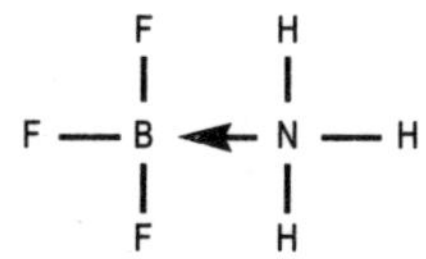

Figure 3.7: Co-ordinate bond in $F_3B \cdot NH_3$

3.6 VSEPR theory and shape of molecules

The shape of covalent molecules can be predicted knowing the fact that the electron pairs (both bonding pairs and lone pairs) arrange themselves around the central atom as far apart as possible so that there is minimum electron repulsion. This is known as Valence Shell Electron Pair Repulsion theory (VSEPR theory).

In molecules of $BeCl_2$, there are two bonding pairs of electrons surrounding Be and they are at an angle of 180° to each other; so $BeCl_2$ is linear. BF_3 is trigonal planar as the three bonding pairs of electrons are at 120°. Similarly CH_4 is tetrahedral as the four bonding electron pairs are at a maximum distance when they are arranged tetrahedrally, and the bond angle in methane is 109.5°. The structures of $BeCl_2$, BF_3 and CH_4 are shown in Figure 3.8.

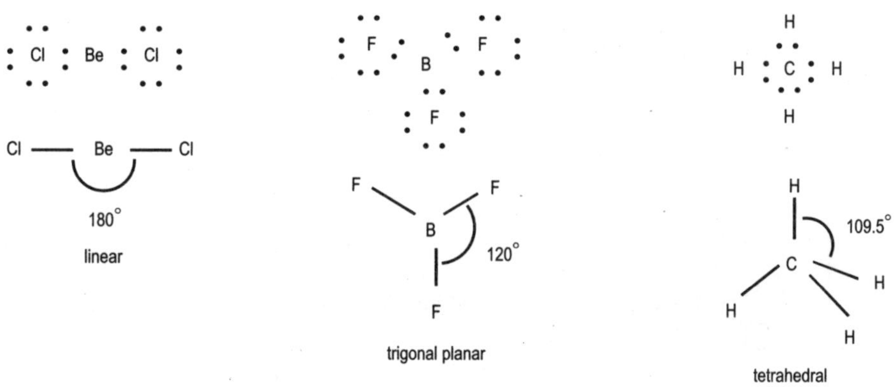

Figure 3.8: Shapes of some covalent molecules

In molecules like H_2O and NH_3, the lone pairs and the bonding pairs of electrons repel each other in order to be at a maximum distance and stability. In these molecules, four electron pairs are arranged tetrahedrally and the bond angle is close to the value for methane. Note that the lone pair-lone pair repulsion is stronger than lone pair-shared pair or shared pair-shared pair repulsions. This explains why the bond angles in water and ammonia are slightly different from what is expected of a regular tetrahedral arrangement.

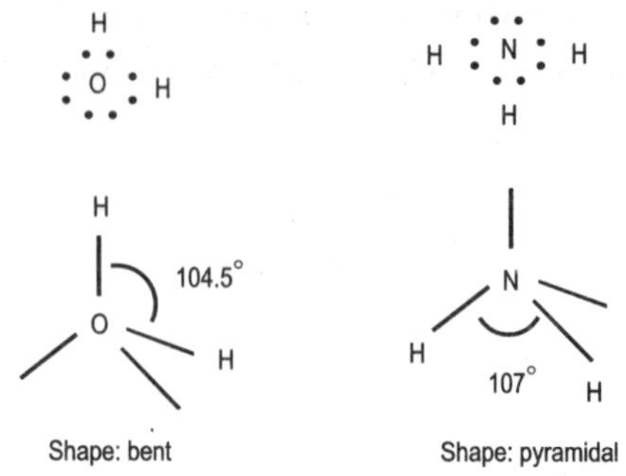

Figure 3.9

Table 3.6

Shape	Description of shape	Examples
	Linear	CO_2
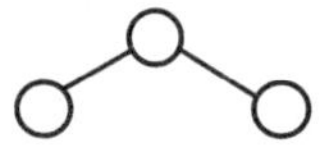	Angular or bent	H_2O, NO_2
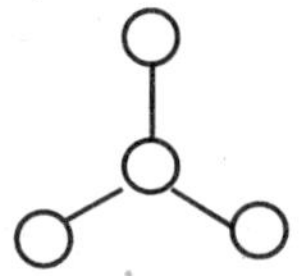	Trigonal planar	BF_3, SO_3
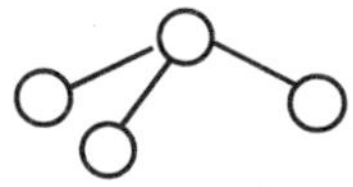	Trigonal pyramidal	NH_3, PCl_3
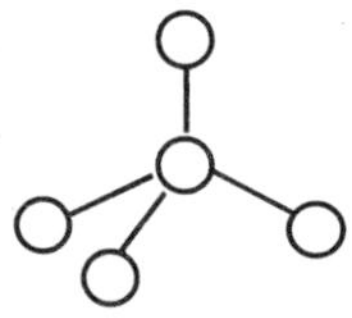	Tetrahedral	CH_4, $SiCl_4$
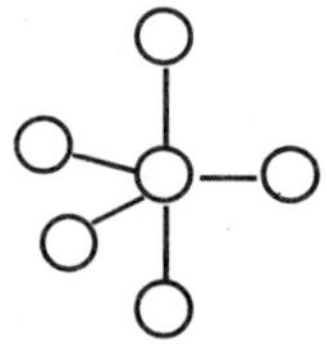	Trigonal bipyramidal	PCl_5
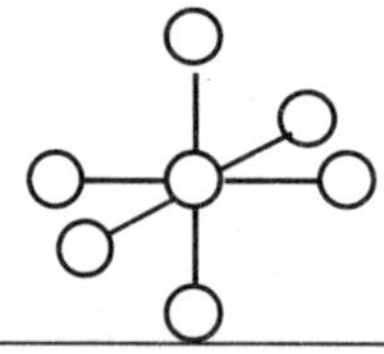	Octahedral	SF_6

Note that the shape of a water molecule is described as 'bent' and that of ammonia as trigonal pyramidal, and not tetrahedral. This is because *the lone pairs are discounted when describing the shape of a molecule.*

3.7 Electronegativity and bond polarity

***Electronegativity** is a measure of the power of an atom to attract the shared electron pair in a covalent bond.* Electronegativity values of elements are calculated using ionisation energies and electron affinities. Electronegativity increases along a Period and decreases down a Group. Note that fluorine is the most electronegative element. Electronegativity values of some elements are given in Table 3.7

The shared electron pair in a covalent bond between two atoms of the *same* element is attracted by both atoms with the same strength. Such a covalent bond is said to be **nonpolar.** H_2 and Cl_2 are examples of nonpolar covalent molecules.

Table 3.7: Electronegativities of some elements

Period 1	H 2.1							He
Period 2	Li 1.0	Be 1.5	B 2.0	C 2.5	N 3.0	O 3.5	F 4.0	Ne
Period 3	Na 0.9	Mg 1.2	Al 1.5	Si 1.8	P 2.1	S 2.5	Cl 3.0	Ar
Period 4							Br 2.8	Kr
Period 5							I 2.5	Xe

An electron pair shared between two atoms of different electronegativities is not shared equally. The more electronegative element attracts the shared electron pair more strongly towards it and such a covalent bond is called a **polar covalent bond.** A compound in which the electron pair is shared unequally is called a **polar covalent compound.**

H—Cl is a polar covalent compound. In H-Cl, the more electronegative Cl atom attracts the shared electron pair closer to it. This shifting of the electron cloud causes the Cl end of the H—Cl molecule to be slightly more negative than the H end. So the Cl end of the molecule becomes partially negatively charged (noted by δ–), and H end, partially positively charged (noted by δ+). The greater the electronegativity difference between the atoms, the higher the bond polarity.

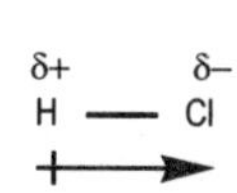

Figure 3.10: Demonstrating the bond polarity in HCl

The bonds in H_2O, CO_2, SO_2 *etc* are polar covalent bonds. The polarity of the whole molecule depends on the polarity of the bonds and the geometry of the molecule.

The **dipole moment** of a molecule gives a measure of the resultant polarity of all the bonds in a molecule. Bond moment is a vector quantity and it has both magnitude and direction. Since an H_2O molecule is bent, the bond moments reinforce each other and the molecule has a resultant dipole moment. So is the case with an SO_2 molecule (Fig. 3.11). Since a CO_2 molecule is linear the resultant dipole moment is zero as bond moments are equal in magnitude but opposite in directions. H_2O and SO_2 are dipolar molecules while CO_2 is nonpolar as a whole. (The dipole moment is represented by an arrow along the bond, pointing from δ+ to δ–).

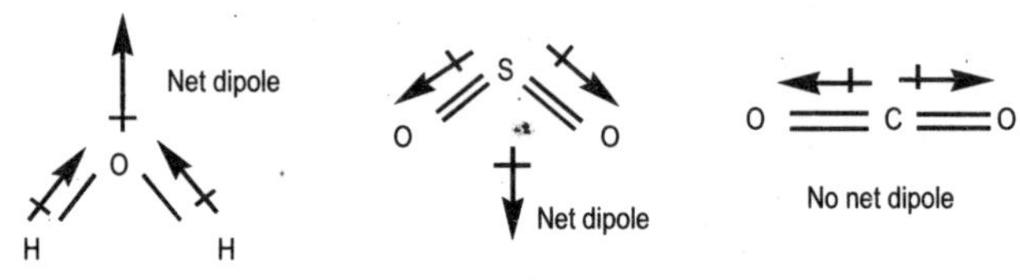

Figure 3.11:

Example 3.8 Which of the following compounds are dipolar? (1) CH_4 (2) NH_3 Give a brief explanation.

Solution

(1) Each of the C-H bonds in CH_4 is polar. The molecule as a whole is nonpolar because of the symmetrical distribution of bonding electron pairs around the carbon atom so that the polarity on one side is balanced by the polarity on the other side.

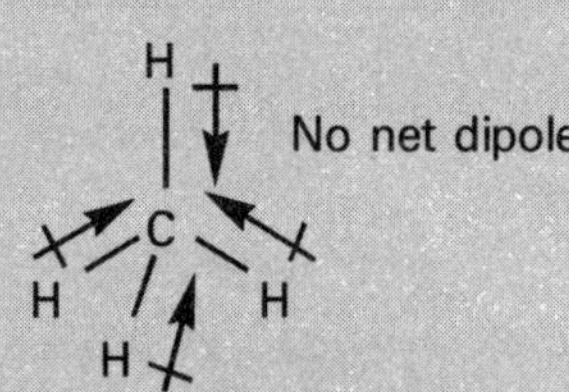

Tetrahedral

(2) Each of the N-H bonds is polar and the molecule as a whole is polar because it has a pyramidal shape.

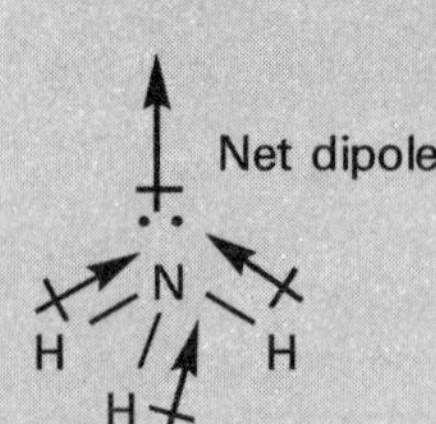

Pyramidal

3.8 Polarization and polarizability

A cation can attract the electron cloud of an anion. Consequently, the electron cloud is distorted. A cation with a high charge density (i.e., small size and high positive charge) has high polarizing power, and an anion which is large and electron rich has high polarizability. As a result of polarization, a covalent character (partial sharing of electrons) develops between the cation and anion.

Let us consider a few examples. Both Li^+ and Na^+ carry one positive charge. A Li^+ is much smaller than a Na^+. So Li^+ has a higher charge density and a higher polarizing power. LiCl and NaCl are both ionic compounds, but they have some covalent character. LiCl is more covalent than NaCl because Li^+ has higher polarizing power. In other words, NaCl is more ionic than LiCl.

In the following series, the covalent character decreases, i.e., the ionic character increases.

$$\underset{\text{Covalent character decreases}}{\underrightarrow{LiCl > NaCl > KCl > RbCl}}$$

$$\underset{\text{Ionic character increases}}{\underrightarrow{LiCl < NaCl < KCl < RbCl}}$$

Let us compare the ionic-covalent character of NaCl and $MgCl_2$. Though Na^+ and Mg^{2+} contain the same number of electrons, Mg^{2+} is smaller than Na^+. Also, Mg^{2+} has two positive charges. The charge density of Mg^{2+} is thus much higher than that of Na^+, its polarizing power is higher and $MgCl_2$ is more covalent than NaCl.

A larger anion is more polarizable than a smaller anion. For example, a Br^- ion is more polarizable than a Cl^- ion. If we compare the covalent nature of NaCl and NaBr, we can see that NaBr is more covalent than NaCl because Br^- is more polarized. Remember that the electronegativity difference between Na and Cl is higher than that between Na and Br and this is another reason for NaCl to be more ionic.

Practice question 3.2 Arrange the silver halides, AgCl, AgBr and AgI, in order of increasing covalent character and give an explanation for the order you have given.

3.9 Continuity of bond types

H_2 and Cl_2 are of nonpolar molecules and can be considered as pure covalent molecules. Compounds like HCl and SO_2 are covalent but have some ionic character due to the polarity of the bonds. Compounds like KCl and KBr are ionic. Even pure covalent molecules can develop a degree of ionic nature (through van der Waals forces) and ionic compounds, some covalent nature (through polarization).

H_2 and Cl_2	**HCl and SO_2**	**KCl and KBr**
Covalent with little ionic character through van der Waals forces	Polar covalent with covalent and ionic nature	Ionic with little covalent nature through polarization

Table 3.8

	H-H	Cl-Cl	H-Cl	NaCl
EN	2.1–2.1	3.0–3.0	2.1–3.0	0.9–3.0
EN difference:	0	0	0.9	2.1

One way of predicting ionic/covalent nature is by using the electronegativity values of atoms. If the difference in electronegativity between atoms in a molecule is less than 1.8, it has a predominantly covalent character, and if above 1.8, ionic character.

HCl is predominantly covalent with some ionic nature and NaCl ionic with some covalent nature.

3.10 Metallic bonding

Metallic bonding can be explained using sodium metal as an example. A sodium atom ($1s^2\ 2s^2 2p^6\ 3s^1$) has one electron in its outermost orbit. When atoms of sodium are arranged closely, the nucleus of an atom attracts its outer shell electron, as well as the outer shell electrons of the nearby atoms. These outer shell electrons move out of their orbits into the space between the positive ions (atomic kernels). These electrons are called **delocalised electrons**. The electrostatic force of attraction between the 'sea' of delocalised electrons and the positive atomic kernels is called **metallic bonding**.

Metallic bonding is a strong type of bonding. If there is more than one electron in the outer orbit of a metal atom, more electrons are delocalised which increases the strength of the metallic bond. For example, a magnesium atom ($1s^2\ 2s^2 2p^6\ 3s^2$) has two outer shell electrons and when these electrons delocalise, a thicker electron cloud is formed. A magnesium ion (Mg^{2+}) also has a double positive charge. The metallic bonding in magnesium is thus stronger than sodium.

Many characteristic properties of metals can be explained in terms of metallic bonding.

1. Metals are good conductors of electricity and heat. Electric current is a flow of electrons. When a metal piece is connected to a supply of electricity, electrons enter the metal and an equal number of electrons from the mobile electron cloud move out, thus conducting electricity.

   When a piece of metal is heated, the positive ions and mobile electrons absorb heat energy and vibrate, making nearby atomic kernels vibrate. The mobile electrons then travel to other parts of the metal carrying this energy to other places. Thus heat is conducted quickly.
2. Metals have a high density. This is due to the fact that strong metallic bonding allows the closest possible packing of atoms.
3. Metals have high melting and boiling points and high heats of fusion and vaporisation. This again can be explained in terms of strong metallic bonds. Large amount of heat energy are required to separate the particles i.e., to break the bonds.

4. Metals are malleable and ductile, i.e. metals can be beaten into different shapes or drawn into wires. When force is applied to a piece of metal, the atomic kernels at the point the metal was struck slide over to other places where they are again held by the attractive forces of the moving electron cloud.
5. Metals are generally lustrous. When light falls on a metal, the electrons absorb light energy and emit most of the energy back.

3.11 Forces between molecules and states of matter

Matter exists in three states, namely; solid, liquid and gas. In the gaseous state, molecules move randomly in all directions, exerting very little force of attraction for one another. At lower temperatures, the kinetic energy of the molecules is low, they become lethargic, intermolecular forces become stronger, and a gas changes to a liquid. In the liquid, the molecules attract one another, are free to move about in the body of the liquid, forming and breaking intermolecular forces. In a solid, molecules are kept at lattice points and they are not free to move about, though vibrational movements take place. In this Section we will study, the different types of forces of attraction in the liquid and solid states.

Ionic solids

We have seen that ionic compounds are solids with high melting and boiling points and ions are held in a lattice by strong electrostatic forces of attraction. For example, sodium chloride solid has a face-centred cubic (FCC) lattice where, if a sodium ion is at the centre of a face of a cube, the chloride ions are at the four corners of the face. In a layer, a sodium ion is surrounded by four chloride ions (almost touching) and in the layers above and below, two chloride ions. So each sodium ion is surrounded by six chloride ions and each chloride ion by six sodium ions. This arrangement repeats in the crystal. A unit cell of sodium chloride is pictured in Figure 3.12.

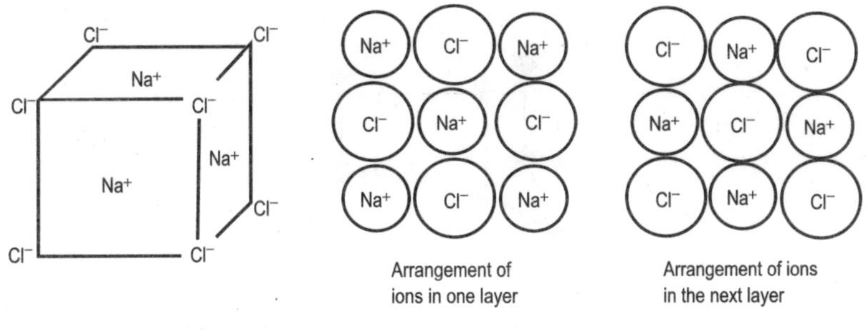

Figure 3.12: Sodium chloride lattice (FCC)

The **co-ordination number** of an ion is the number of oppositely charged ions surrounding it. The co-ordination number of sodium and chlorine in sodium chloride is each six.

Depending on the size of the ions and the number of ions in the compound, there can be other arrangements. Caesium chloride has a body-centred cubic (BCC) lattice. If a caesium ion is at the centre of a cube, there are chloride ions at the eight corners of

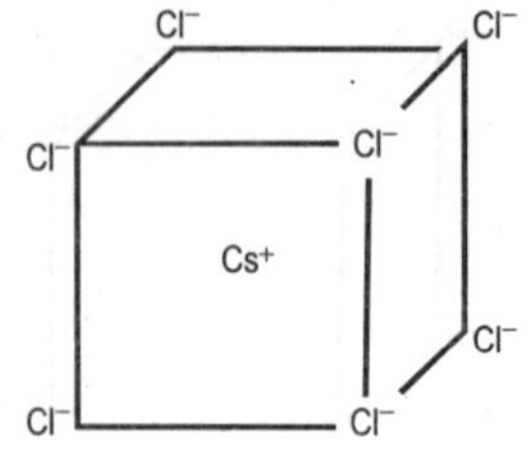

Figure 3.13: Caesium chloride lattice (BCC)

the cube and *vice versa*. The co-ordination number of caesium and chlorine in caesium chloride is eight.

Metallic solids

Metals form crystalline lattices where atomic kernels are arranged as closely as possible, such as hexagonal cubic packing (HCP) and body centred cubic packing (BCC).

Network solids

Some non-metallic elements like carbon and silicon are solids with high melting points. In diamond, an allotrope of carbon, each carbon atom is bonded to four other carbon atoms tetrahedrally by strong covalent bonds. This arrangement repeats in the crystal (Figure 3.14). Silicon has a lattice similar to diamond. Graphite, which is another allotrope of carbon has a layer lattice. In a layer, each carbon atom is bonded to three other carbon atoms by strong covalent bonds (Figure 3.15), while there are van der Waals forces of attraction (explained below) between the layers. The high melting point of such a substance is due to the large amount of energy required to break all these covalent bonds.

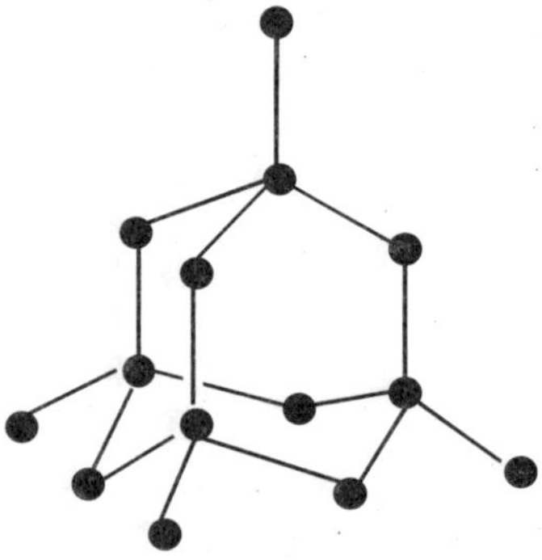

Figure 3.14: The tetrahedral arrangement of carbon atoms in Diamond

Other types of bonding

There are other types of bonds present between atoms or molecules in the liquid and solid state. For example, the forces of attraction holding water molecules in water and ice are **hydrogen bonds**. The type of forces between argon atoms in liquid argon, and between iodine molecules in solid iodine are **van der Waals forces**, that between sulphur dioxide molecules is **dipole-dipole attraction**. These are discussed below.

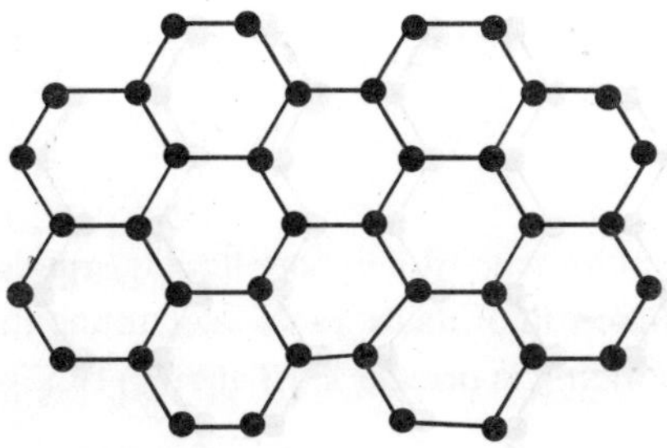

Figure 3.15: A layer of carbon atoms in graphite

Hydrogen bonding

The main force of attraction between molecules which contain a hydrogen atom bonded to a more electronegative element like O and Cl, is *hydrogen bonding*. Water is a polar covalent compound having a δ +ve H and δ –ve O in the molecule. The δ +ve

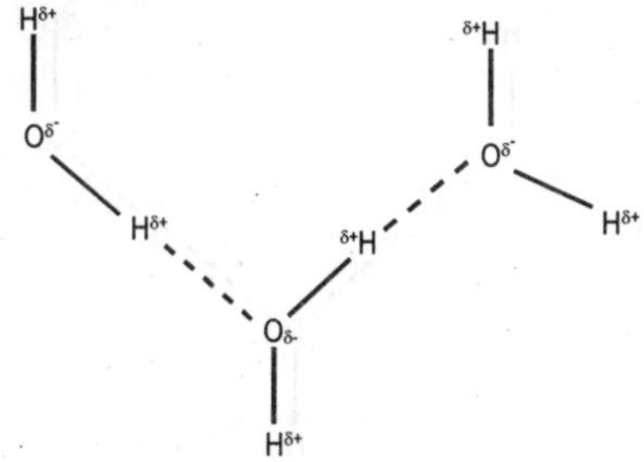

Hydrogen bonding between water molecules, shown by dotted lines

Figure 3.16

H end of a water molecule attracts the δ –ve O end of another molecule through the lone pair of electrons on the oxygen. This type of attractive force is called **hydrogen bonding.**

Other examples of substances which exhibit hydrogen bonding in the liquid or solid states are hydrogen halides (HF, HCl, HBr and HI), ammonia and hydrogen sulphide. The essential conditions for hydrogen bond formation between molecules are

- hydrogen atom bonded to a more electronegative element
- lone pair(s) of electrons on the more electronegative element.

Methane (CH_4) molecules are not hydrogen bonded. Though each C-H bond is polar, there are no lone pairs of electrons on carbon. Water is unique in the sense that O has two lone pairs of electrons and there are two δ +ve H atoms per molecule so that each water molecule forms four hydrogen bonds with other water molecules, two through H atoms and two through the lone pairs of electrons on O. This explains the high melting and boiling points of water compared with say, hydrogen fluoride and ammonia.

Example 3.9 Which of the two has a higher boiling point, HF or HCl? Give an explanation for your answer.

Solution
HF has a higher boiling point than HCl. In both liquids, molecules are held together by hydrogen bonds. Since fluorine is more electronegative than chlorine, the strength of hydrogen bonding is greater in HF than in HCl.

Van der Waals forces or London forces

Since electron clouds around nuclei in molecules are not rigid and can fluctuate, a non-polar molecule can become a dipole at any moment when the centre of the positive charge does not coincide with the centre of the negative charge. This is called an *instantaneous dipole*. Such a dipole can polarise a neighbouring molecule (induced dipole) and a force of attraction is developed between them. This type of attraction between instantaneous dipoles is termed **van der Waals forces.** Instantaneous dipoles are formed and 'die' continuously.

The forces of attraction between atoms like He or molecules like H_2 and CH_4 in liquid or solid states are van der Waals forces. Van der Waals forces are weak forces of attraction, the strength of which increases with electron density which in turn depends on molecular weight.

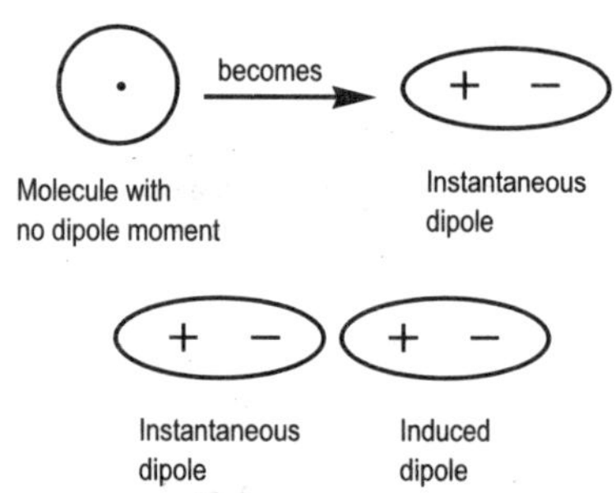

Figure 3.17

Example 3.10 Which of the two has a higher boiling point, methane (CH_4) or ethane (C_2H_6)?

Solution
The molecular weight of ethane is higher, the van der Waals forces between ethane molecules are stronger and so, ethane has a higher boiling point than methane.

Practice question 3.3
The boiling points of the noble gases, in degrees Kelvin, are; He, 4, Ne, 27, Ar, 87, Kr, 121 and Xe, 166. Give an explanation for this trend.

Dipole-dipole attraction

Polar covalent molecules like iodine monochloride (ICl) or sulphur dioxide (SO_2) have δ+ve and δ–ve ends in each molecule. They are called permanent dipoles. The oppositely charged ends of two dipoles attract. This type of force of attraction between two dipoles is called **dipole-dipole attraction**. Dipole-dipole attractions between molecules are in addition to van der Waals forces.

The strength of dipole-dipole attractions depends on the polarity of the bonds and the shape of the molecule (Section 3.6).

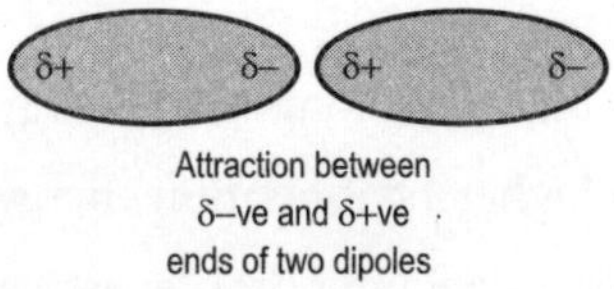

Figure 3.18

Tutorial: helping you learn

Progress questions

3.1 The 1st ionisation energy of lithium is +520 kJ mol^{-1} and its 2nd ionization energy is +7298 kJ mol^{-1}. Explain why the 2nd ionisation energy is much higher than the 1st.

3.2 Which of the two elements has a higher first ionisation energy, Na ($Z = 11$) or Mg ($Z = 12$)? Give an explanation for your answer.

3.3 Which of the two elements has a higher second ionisation energy, Na or Mg? Why?

3.4 Which of the two has a higher 1st ionisation energy, Mg or Al? Why?

3.5 Which of the two has a higher 1st ionisation energy, Na or K? Why?

3.6 Given below are the successive ionisation energies of sodium. Plot a graph of the log of the ionisation energies against the number of electrons, and give an explanation for the trend and any breaks in the graph.

Table 3.9: Ionisation energies in kJ mol^{-1}

1	2	3	4	5	6	7	8	9	10	11
496	4563	6913	9544	13352	16611	20115	25491	28934	141367	159079

Successive ionisation energies of sodium

3.7 How does the first ionisation energy vary in the noble gas family; helium, neon, argon, krypton and xenon, with increase in atomic number? Give reasons for the trend.

3.8 Write (a) the electronic configuration, and (b) the Lewis electron-dot structure of these elements
(i) B ($Z = 5$) (ii) Si ($Z = 14$) (iii) S ($Z = 16$) (iv) Cl ($Z = 17$)

3.9 Predict the shape of the molecules
(a) H_2S (b) CS_2 (c) BCl_3 (d) SCl_6 (e) CCl_4 (f) PH_3

3.10 Group the following into ionic and covalent compounds
(a) $NaNO_2$ (b) CaO (c) NO_2 (d) PCl_5 (e) SiO_2 (f) $MgBr_2$
(g) $CuSO_4$ (h) ClO_2

3.11 Which has a greater covalent nature; AgF or AgCl? Why?

3.12 Arrange the following groups of compounds in the order of increasing polarity
(a) AgCl, AgBr, AgI (b) HF, HCl, HBr (c) BeO, MgO, CaO

3.13 For element 11, name and describe the type of bonding you would expect to be formed between its atoms in the solid state.

4 Energy Changes

One-minute summary

Chemical reactions are almost always accompanied by energy changes (in the form of heat and light). Heat energy is either evolved or absorbed during a chemical reaction. The amount of heat evolved or absorbed depends on the quantities of substances that take part in the reaction. Chemists use the term 'enthalpy' to describe the amount of transferable heat. In this chapter we will discuss enthalpy and enthalpy changes during reactions and various methods of calculation of enthalpy changes. Hess's law of constant heat summation and its applications are explained.

The topics covered include:

- exothermic and endothermic reactions
- enthalpy of formation
- enthalpy of combustion
- enthalpy diagrams
- Hess's law
- calculation of enthalpy of reaction
- bond enthalpy
- experimental determination of enthalpy of a reaction

4.1 Important terms and definitions

Chemical energy is the form of energy stored in chemical substances, and it depends on the type and arrangements of atoms and molecules. Almost all chemical reactions are accompanied by absorption or evolution of energy; generally in the form of heat energy.

Thermochemistry is the study of heat changes in chemical reactions.

An **exothermic reaction** is one which gives out heat to the surroundings. An example of an exothermic reaction is the formation of water from hydrogen and oxygen (Figure 4.1).

An **endothermic reaction** is accompanied by the absorption of heat. When carbon combines with hydrogen to form ethyne (acetylene) heat is absorbed and this is an endothermic reaction (Figure 4.2).

Enthalpy, represented by the letter ***H***, is the term used by chemists to express the quantity of heat transferred at constant pressure during a reaction.

Enthalpy of a reaction or **enthalpy change, *ΔH*** (delta *H*), is the difference between the enthalpies of products and the enthalpies of reactants.

$$\Delta H = H(\text{products}) - H(\text{reactants})$$

Since heat is given out during an exothermic reaction, the sum of the enthalpies of the products is lower than that of the reactants. So ΔH is ***negative*** for an exothermic reaction. ΔH is ***positive*** for an endothermic reaction since heat is absorbed by the system and the sum enthalpies of the products is higher than that of the reactants.

A **thermochemical equation** is a stoichiometric equation for a reaction which also includes the enthalpy change for the quantities given by the equation. For example,

$$2H_2(g) + O_2(g) \longrightarrow 2H_2O(l), \ \Delta H = -571.6 \text{ kJ}$$

This thermochemical equation informs us that 571.6 kJ of heat is produced (ΔH negative) when 2 moles of hydrogen and 1 mole of oxygen react completely to form two moles of water. Or, we can say that 285.8 kJ of heat is produced, during the formation of 1 mole of water from hydrogen and oxygen. It is important to show the states of the reactants and products, using (g) for a gas, (l) for a liquid, (s) for a solid and (aq) for aqueous.

Standard conditions The enthalpy change of a reaction can vary with changes in temperature and pressure. So **standard conditions** must be used for the comparison of enthalpy changes. **Standard pressure is 1 atmosphere (1 atm) and standard temperature is 298 K (25° C)**. The symbol ΔH° is used to depict the enthalpy change at standard conditions.

An **enthalpy diagram** portrays the *relative levels* of the enthalpies of the reactants and products, and the *enthalpy change* during a reaction. Figure 4.1 shows the enthalpy diagram for the formation of water.

$$2H_2(g) + O_2(g) \longrightarrow 2H_2O(l), \quad \Delta H = -571.6 \text{ kJ}$$

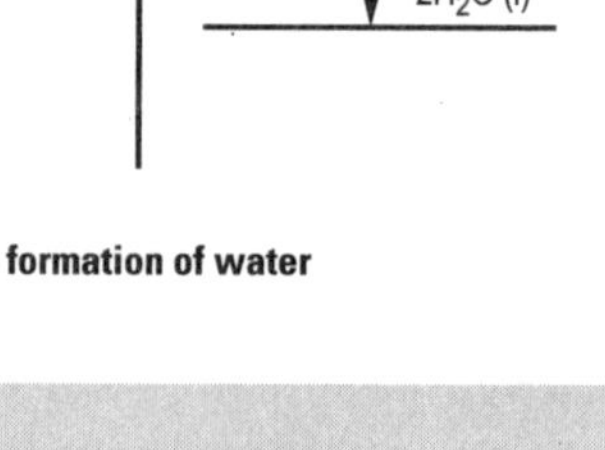

Figure 4.1: Enthalpy diagram for the formation of water

> **Notes**
> 1. The formation of H_2O from H_2 and O_2 is an exothermic reaction. The enthalpy change for an exothermic reaction is shown by a downward arrow.
> 2. 571.6 kJ heat is evolved when 2 moles of H_2O are formed from 2 moles of H_2 and 1 mole of O_2.
> 3. The enthalpy of H_2O is lower than the sum of the enthalpies of H_2 and O_2.

$$2C(s) + H_2(g) \longrightarrow C_2H_2(g), \ \Delta H = +228.0 \text{kJ}$$

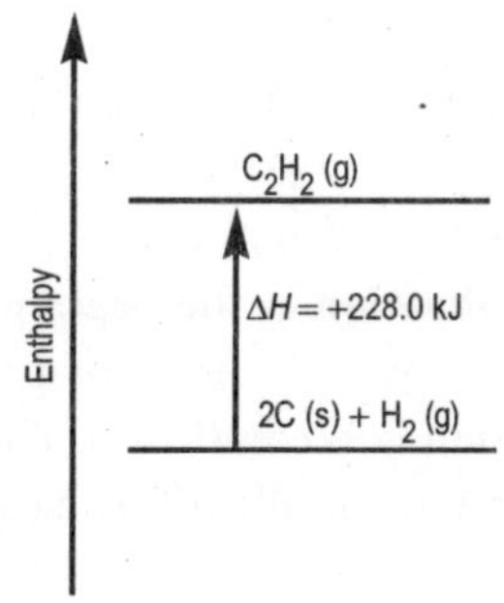

Figure 4.2: Enthalpy diagram for an endothermic reaction

Notes

1. Formation of ethyne (C_2H_2) is an endothermic reaction. The enthalpy change for an endothermic reaction is shown by an upward arrow.
2. 228.0 kJ of heat is absorbed during the formation of one mole of C_2H_2 gas from 2 moles of C and 1 mole of H_2.
3. The enthalpy of ethyne is higher than the sum of the enthalpies of C and H_2.

4.2 Standard enthalpy changes of reactions

Standard enthalpy change of formation, ΔH°_f, is the heat *evolved* or *absorbed* when 1 mole of a substance is formed from its constituent elements that is, 1 atm pressure and 298 K.

The standard enthalpies of formation of all *elements* are assigned 0 kJ mol^{-1} at 1 atm. and 298 K and that of a compound is either positive or negative, depending on whether the compound is formed from its constituent elements by an endothermic or exothermic reaction. Some examples of elements in their standard states are $H_2(g)$, $O_2(g)$, C (graphite), Fe(s), Na(s) and Al(s).

The enthalpy of formation, ΔH°_f, of water is –285.8 kJ mol^{-1}. That is, 285.8 kJ of heat is evolved when 1 mole of water is formed from hydrogen and oxygen gases at standard conditions. The ΔH°_f, of methane is –74.8 kJ mol^{-1}. The thermochemical equations for the enthalpies of formation of water and methane are given below.

$$H_2(g) + \frac{1}{2}O_2(g) \longrightarrow H_2O(l), \ \Delta H^\circ_f = -285.8 \text{ kJ mol}^{-1}$$

$$C(s) + 2H_2(g) \longrightarrow CH_4(g), \ \Delta H^\circ_f = -74.8 \text{ kJ mol}^{-1}$$

Standard enthalpy of combustion, ΔH°_c, is the heat *evolved* when 1 mole of a substance is burnt completely in air or oxygen at standard conditions, that is, 1 atm pressure and 298 K. Note that combustion reactions are always exothermic.

For example, the enthalpy of combustion of methane is –890.3 kJ mol^{-1} which means, 890.3 kJ of heat is produced when 1 mole of methane is converted to carbon dioxide and water.

$$CH_4(g) + 2O_2(g) \longrightarrow CO_2(g) + 2H_2O(l), \ \Delta H = -890.3 \text{ kJ mol}^{-1}$$

Standard enthalpy of solution is the heat *evolved* or *absorbed* when 1 mole of a substance is dissolved in excess water so that further dilution causes no heat change. Some substances dissolve exothermically and some endothermically. For example, when 1 mole of sodium chloride dissolves in excess water, 3.9 kJ of heat is absorbed.

$$NaCl(s) + aq \longrightarrow Na^+(aq) + Cl^-(aq), \quad \Delta H = +3.9 \text{ kJ mol}^{-1}$$

Standard enthalpy of neutralisation is the heat *evolved* when 1 mole of $H_2O(l)$ is formed by the reaction of H^+ ions and OH^- ions provided by a dilute acid and a dilute alkali respectively under standard conditions. The enthalpy of neutralisation is constant for any strong acid with a strong alkali. Strong acids and alkalis are completely dissociated in dilute solutions.

$$HCl(g) + aq \longrightarrow H^+(aq) + Cl^-(aq)$$
$$NaOH(s) + aq \longrightarrow Na^+(aq) + OH^-(aq)$$

When an acid is mixed with an alkali, the H^+ from the acid reacts with the OH^- ions from the alkali to form H_2O and this is neutralisation. The enthalpy of neutralisation is given by the equation,

$$H^+(aq) + OH^-(aq) \longrightarrow H_2O(l), \quad \Delta H = -57.6 \text{ kJ mol}^{-1}$$

The enthalpy of neutralisation of a weak acid or weak base is different. A weak acid or base does not dissociate completely in solution and heat is required for the complete dissociation so that part of the heat that would have evolved during the neutralisation is used up for the dissociation. So the net heat change during the neutralisation with a weak acid or base is less.

Standard enthalpy of a reaction, in general, is the heat evolved or absorbed when the number of moles of the reactants given by a stoichiometric equation react completely under standard conditions of 1 atm and 298 K. For example,

$$2H_2S(g) + SO_2(g) \longrightarrow 3S(s) + 2H_2O(l), \quad \Delta H = -233.6 \text{ kJ}$$

In the above reaction, 233.6 kJ of heat is evolved when 2 moles of hydrogen sulphide react with 1 mole of sulphur dioxide to form the products under standard conditions.

Example 4.1 Calculate the heat evolved when 5.0 g of hydrogen react completely with oxygen to form water under standard conditions. (A_r: H = 1)

$$2H_2(g) + O_2(g) \longrightarrow 2H_2O(l), \quad \Delta H = -571.6 \text{ kJ}$$

Solution

$$\text{Number of moles of } H_2 = \frac{\text{Mass}}{\text{Molar mass}} = \frac{5.0 \text{ g}}{2 \text{ g mol}^{-1}} = 2.5 \text{ mol}$$

2 moles of H_2 produce 571.6 kJ of heat

$$2.5 \text{ moles of } H_2 \text{ produce } \frac{571.6 \times 2.5}{2} = 714.5 \text{ kJ of heat}$$

Example 4.2 Butane (C_4H_{10}) burns in air to give carbon dioxide and water. Given that the enthalpy of combustion of butane is –2876.5 kJ mol^{-1}, calculate the heat change during the combustion of 20.0 g of butane. (A_r : H = 1, C = 12)

Solution

$$\text{Molar mass of butane } (C_4H_{10}) = 58 \text{ g mol}^{-1}$$

$$\text{Number of moles of butane} = \frac{\text{Mass}}{\text{Molar mass}} = \frac{20.0 \text{ g}}{58 \text{ g mol}^{-1}} = 0.345 \text{ mol}$$

Enthalpy of combustion of butane = –2876.5 kJ mol^{-1}
i.e., 1 mol of butane produces 2876.5 kJ heat on combustion
0.345 mol of butane produces 2876.5 × 0.345 = 992.4 kJ heat

4.3 Hess's Law and its applications

Hess's Law states that *whether a reaction takes place in one step or in a series of steps, the total heat change remains the same.* For example, carbon dioxide can be formed from carbon and oxygen by a direct reaction (1) or through the formation of carbon monoxide (2 and 3), as shown below. The heat of formation of carbon dioxide is –393.5 kJ mol^{-1}, whether it is formed in one or two steps.

$$C(s) + O_2(g) \longrightarrow CO_2(g),\ \Delta H_1 = -393.5 \text{ kJ mol}^{-1} \quad (1)$$

$$C(s) + \frac{1}{2}O_2(g) \longrightarrow CO(g),\ \Delta H_2 = -110.5 \text{ kJ mol}^{-1} \quad (2)$$

$$CO(g) + \frac{1}{2}O_2(g) \longrightarrow CO_2(g),\ \Delta H_3 = -283.0 \text{ kJ mol}^{-1} \quad (3)$$

Add equations (2) and (3).

$$C(s) + \cancel{CO}(g) + O_2(g) \longrightarrow \cancel{CO}(g) + CO_2(g),$$

$$\Delta H = -110.5 + -283.5 = -393.5 \text{ kJ mol}^{-1}$$

$$\text{i.e., } C(s) + O_2(g) \longrightarrow CO_2(g),\ \Delta H = -393.5 \text{ kJ mol}^{-1}$$

These reactions are shown in the following enthalpy diagram. Note that 1 mol of C combines with 1 mol of O_2 to form CO_2 in one step. In the alternate route, 1 mol of C combines with 1/2 mol of O_2 to form CO which combine with the other 1/2 mol of O_2 to form CO_2. According to Hess's law, $\Delta H_1 = \Delta H_2 + \Delta H_3$ and this can be noticed from the diagram (Figure 4.3).

The important application of Hess's law is that thermochemical equations can be added, subtracted or multiplied by a factor, like algebraic equations. This enables us to calculate enthalpy changes of reactions using enthalpies of formation or using other enthalpies of reactions. This is illustrated in Example 4.6

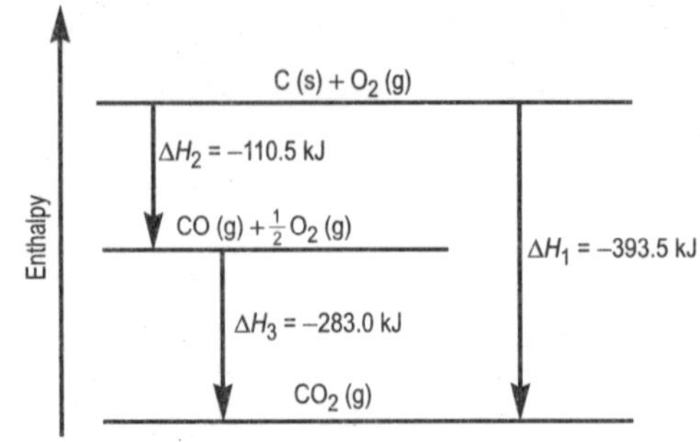

Figure 4.3: Enthalpy diagram for the formation of carbon dioxide

4.4 Calculation of enthalpy of a reaction using standard enthalpies of formation

The enthalpy of a reaction can be calculated from the standard enthalpies of formation of the reactants and products, using the relation,

$$\Delta H_{rxn} = \Sigma\, \Delta H^{o}_{f\,(products)} - \Sigma\, \Delta H^{o}_{f\,(reactants)}$$

(Σ stands for sum of the enthalpies.)

Example 4.3 Calculation of enthalpy of reaction using ΔH^{o}_{f}

Iron (III) oxide reacts with aluminium according to the equation,

$$Fe_2O_3(s) + 2Al(s) \longrightarrow Al_2O_3(s) + 2Fe(s)$$

Calculate the enthalpy of reaction, given the standard enthalpies of formation.

Substance	**ΔH^{o}_{f} (kJ mol^{-1})**
Al(s)	0
Fe(s)	0
$Fe_2O_3(s)$	–824.2
$Al_2O_3(s)$	–1675.7

Solution

$$Fe_2O_3(s) + 2Al(s) \longrightarrow Al_2O_3(s) + 2Fe(s)$$

$$\Delta H_{rxn} = \Sigma\, \Delta H^{o}_{f\,(products)} - \Sigma\, \Delta H^{o}_{f\,(reactants)}$$

$$\Delta H_{rxn} = [\Delta H^{o}_{f}\,(Al_2O_3) + 2 \times \Delta H^{o}_{f}\,(Fe)] - [\Delta H^{o}_{f}\,(Fe_2O_3) + 2 \times \Delta H^{o}_{f}\,(Al)]$$

$$= (-1675.7 + 2 \times 0) - (-824.2 + 2 \times 0) = -851.5 \text{ kJ}$$

Example 4.4 Calculate the enthalpy of the reaction

$$CH_4(g) + 2O_2(g) \longrightarrow CO_2(g) + 2H_2O(l)$$

using the standard enthalpies of formation.

Substance	**ΔH^{o}_{f} (kJ mol^{-1})**
CH_4	–74.8
O_2	0
CO_2	–393.5
H_2O	–285.8

Solution

$$\Delta H_{rxn} = [\Delta H^\circ_f (CO_2) + 2 \times \Delta H^\circ_f (H_2O)] - [\Delta H^\circ_f (CH_4) + 2 \times \Delta H^\circ_f (O_2)]$$
$$= [-393.5 + (2 \times -285.8)] - [-74.8 + (2 \times 0)]$$
$$= -890.3 \text{ kJ}$$

4.5 Calculation of enthalpy of a reaction using an enthalpy diagram

In Sections 4.1 and 4.3, we have discussed how to draw an enthalpy diagram for a reaction. An enthalpy diagram for a set of reactions can be drawn as illustrated in Example 4.5. In this example, the enthalpies of formation of methane, carbon dioxide and water are shown in the enthalpy diagram and the enthalpy of combustion of methane is calculated.

Example 4.5 Enthalpy diagram to calculate the enthalpy of a reaction

Draw an enthalpy diagram for the following reactions, and use the diagram to determine the enthalpy of combustion of methane.

$$C(s) + 2H_2(g) \longrightarrow CH_4(g), \ \Delta H_1 = -74.8 \text{ kJ}$$
$$C(s) + O_2(g) \longrightarrow CO_2(g), \ \Delta H_2 = -393.5 \text{ kJ}$$
$$2H_2(g) + O_2(g) \longrightarrow 2H2O(l), \ \Delta H_3 = -571.6 \text{ kJ}$$

Hint: The elements involved in the three reactions are 1 mol of C (s), 2 mol of H_2(g) and 2 mol of O_2(g) and these form the starting level. Then draw the diagram showing the reactions as follows.

Solution

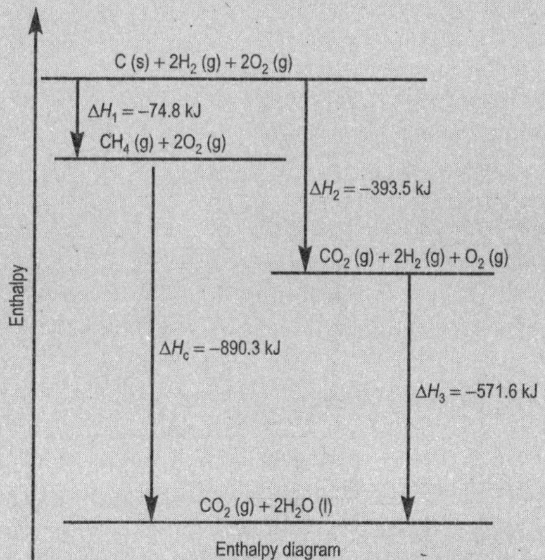

Enthalpy diagram

The equation for the combustion reaction is,

$$CH_4(g) + 2O_2(g) \longrightarrow CO_2(g) + 2H_2O(l)$$

The enthalpy of combustion of methane, ΔH_c, is calculated using the diagram,

$$\Delta H_c = \Delta H_2 + \Delta H_3 - \Delta H_1$$
$$= [-393.5 + (-571.6)] - (-74.8) = -890.3 \text{ kJ}$$

$$CH_4(g) + 2O_2(g) \longrightarrow CO_2(g) + 2H_2O(l), \ \Delta H_c = -890.3 \text{ kJ}$$

Example 4.6 Calculation of enthalpy of a reaction

Calculate the enthalpy of the reaction,

$$PCl_3(l) + Cl_2(g) \longrightarrow PCl_5(s)$$

given the enthalpies of formation of $PCl_3(l)$ and $PCl_5(s)$.

$$P(s) + \frac{3}{2}Cl_2(g) \longrightarrow PCl_5(l), \ \Delta H = -320.0 \text{ kJ mol}^{-1} \quad (1)$$

$$P(s) + \frac{5}{2}Cl_2(g) \longrightarrow PCl_5(s), \ \Delta H = -443.0 \text{ kJ mol}^{-1} \quad (2)$$

Solution: Method 1 (applying Hess's law)

The enthalpy change for the required reaction can be calculated indirectly using Hess's law. Since we need PCl_3 and Cl_2 as the reactants and PCl_5 as the product, reverse equation (1), add the result to equation (2). (Remember that when an equation is reversed, the sign of ΔH is also reversed.)

$$PCl_3(l) \longrightarrow P(s) + \frac{3}{2}Cl_2(g), \ \Delta H = +320.0 \text{ kJ mol}^{-1}$$

$$P(s) + \frac{5}{2}Cl_2(g) \longrightarrow PCl_5(s), \ \Delta H = -443.0 \text{ kJ mol}^{-1}$$

$$PCl_3(l) + P(s) + \frac{5}{2}Cl_2(g) \longrightarrow P(s) + \frac{3}{2}Cl_2(g) + PCl_5(s), \quad \Delta H = -123.0 \text{ kJ mol}^{-1}$$

The result is,

$$PCl_3(l) + Cl_2(g) \longrightarrow PCl_5(s), \ \Delta H = -123.0 \text{ kJ mol}^{-1}$$

Note that when two equations are added, substances which are common to the two sides of the equations are cancelled out.

Method 2 (using an enthalpy diagram)

Alternatively, an enthalpy diagram can be drawn for the reactions given by the equations (1) and (2) (above) and from the diagram the enthalpy change of the required reaction can be worked out directly as –443.0 – (–320.0) = –123.0 kJ.

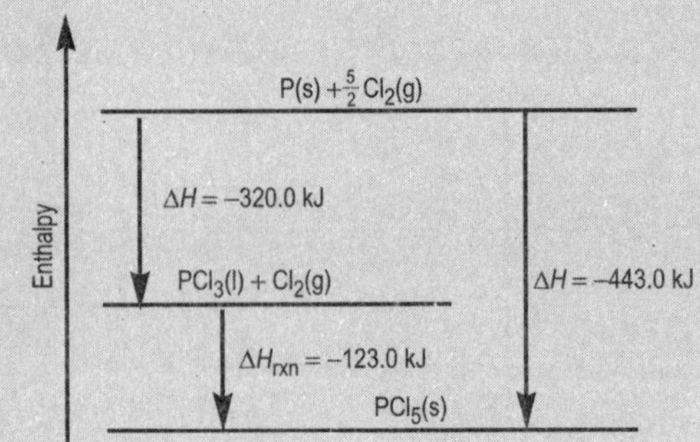

Figure 4.4: Enthalpy diagram for the formation of PCl_5

Method 3 (using enthalpies of formation)
The enthalpy change for the reaction

$$PCl_3(l) + Cl_2(g) \longrightarrow PCl_5(s)$$

can also be calculated using the relation,

$$\Delta H_{rxn} = \Sigma\, \Delta H^{\circ}_{f\,(products)} - \Sigma\, \Delta H^{\circ}_{f\,(reactants)}$$

$$\Delta H_{rxn} = [\Delta H^{\circ}_f\,(PCl_5)] - [\Delta H^{\circ}_f\,(PCl_3) + \Delta H^{\circ}_f\,(Cl_2)]$$

$$= -443.0 - (-320.0 + 0) = -123.0 \text{ kJ mol}^{-1}$$

4.6 Experimental determination of enthalpy changes

Calorimetry

The enthalpy changes of many reactions are determined in the laboratory by carrying out the reactions using convenient amounts of the reactants in a calorimeter and measuring the heat change during reactions. The heat evolved during an exothermic reaction is taken up by the calorimeter and its contents resulting in a rise in temperature of the system. If it is an endothermic reaction, the heat required for the reaction is taken from the calorimeter and its contents, causing a fall in temperature. The change in temperature can be measured using a thermometer.

If a polystyrene cup is used as the calorimeter, the heat absorbed by it is negligibly small and can be ignored. The heat evolved or absorbed can be calculated using the relation,

$$\text{Enthalpy change} = mc\Delta T$$

where m is the mass of the contents of the calorimeter, c is the specific heat capacity of the solution and ΔT the change in temperature.

The specific heat capacity of a substance is the energy required to change the temperature of 1 g of the substance by 1 K. The specific heat capacity of water is 4.18 J g^{-1} K^{-1}. This means that 4.18 Joules of energy is needed to raise the temperature of 1 g of water by 1 degree Kelvin. The specific heat capacity of an aqueous solution is taken as the specific heat capacity of water. From the temperature change and the mass of the contents, enthalpy change can be calculated.

The heat of neutralisation of hydrochloric acid with sodium hydroxide solution can be determined using simple materials available in a laboratory. The following example shows how the heat of a neutralisation reaction is determined.

Example 4.7
50.0 cm^3 of hydrochloric acid of 2.0 mol dm^{-3} is mixed with 50.0 cm^3 of sodium hydroxide solution of the same molar concentration and same initial temperature, in a polystyrene cup. The rise in temperature of the solution is found to be 13.6° C. Calculate the heat of neutralisation of hydrochloric acid, assuming that the heat produced during the reaction is completely absorbed by the solution., and the specific heat capacity and density of the solution are the same as for water. (Specific heat capacity of water = 4.18 J g^{-1} K^{-1} and density of water = 1 g cm^{-3}.)

Solution

The heat absorbed by the solution $= mc\Delta T$

where m is the mass of the solution, c the specific heat capacity of the solution and ΔT the rise in temperature.

$$\text{The volume of the solution} = 50 + 50 = 100\ \text{cm}^3$$

$$\text{Mass of the solution} = \text{Volume} \times \text{density}$$

$$= 100\ \text{cm}^3 \times 1\ \text{g cm}^{-3} = 100\ \text{g}$$

$$\text{The heat absorbed by the solution} = 100\ \text{g} \times 4.18\ \text{J g}^{-1}\ \text{K}^{-1} \times 13.6\ \text{K} = 5726.6\ \text{J}$$

$$\text{Molar concentration of HCl} = 2.0\ \text{mol dm}^{-3}$$

$$\text{Volume of HCl} = \frac{50}{100} = 0.05\ \text{dm}^3$$

$$\text{Number of moles of HCl reacted} = 2.0\ \text{mol dm}^{-3} \times 0.05\ \text{dm}^3 = 0.1\ \text{mol}$$

$$0.1\ \text{mol HCl produces} = 5726.6\ \text{J heat}$$

$$1\ \text{mol HCl produces} = \frac{5726.6 \times 1}{0.1} = 57266\ \text{J}$$

$$= 57.27\ \text{kJ}$$

$$\text{Heat of neutralisation, } \Delta H = -57.27\ \text{kJ mol}^{-1}$$

4.7 Bond enthalpies and enthalpy changes of reactions

Bond enthalpy or **bond dissociation energy** (BDE) is the heat required to break the bonds in 1 mole of gaseous diatomic molecules to form gaseous atoms. For example, the bond enthalpy of hydrogen chloride is given by the following equation,

$$\text{HCl(g)} \longrightarrow \text{H(g)} + \text{Cl(g)},\ \Delta H_{\text{BDE}} = +431.0\ \text{kJ mol}^{-1}$$

1 mole of HCl contains 1 mole of bonds and 431.0 kJ of heat is required to break these bonds.

1 mole of water contains 2 moles of O–H bonds.

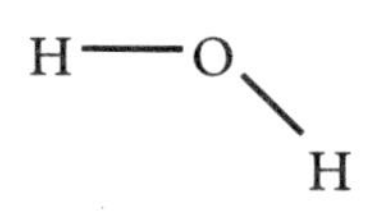

The (average) bond dissociation energy of the O–H bond is half the amount of the heat required to convert 1 mole of water in the gaseous state to gaseous H and O atoms.

$$\text{H}_2\text{O(g)} \longrightarrow 2\text{H(g)} + \text{O(g)},\ \Delta H = +928.0\ \text{kJ mol}^{-1}$$

$$\Delta H_{\text{BDE}}\ (\text{O–H}) = \frac{928.0}{2} = +464.0\ \text{kJ mol}^{-1}$$

The average bond dissociation energies of some covalent bonds are given in Table 4.1

Table 4.1: Bond dissociation energies

Bond	BDE (kJ mol^{-1})	Bond	BDE (kJ mol^{-1})
H–H	435.9	C–H (in CH_4)	435.0
H–Cl	432.0	C=C	612.0
H–Br	366.3	C≡C	838.0
Br–Br	192.9	O=O	498.3
Cl–Cl	243.4	N≡N	945.4
O–H	464.0	C–O	358.0
N–H	391.0	C–Cl	346.0
C–H (average)	413.0	C–C	347.0

For a molecule which contains more than one covalent bond, the total bond dissociation energy is the sum of the bond dissociation energies of all the bonds. For example, a molecule of ethanol contains five C–H bonds, one C–C bond, one C–O bond and one O–H bond.

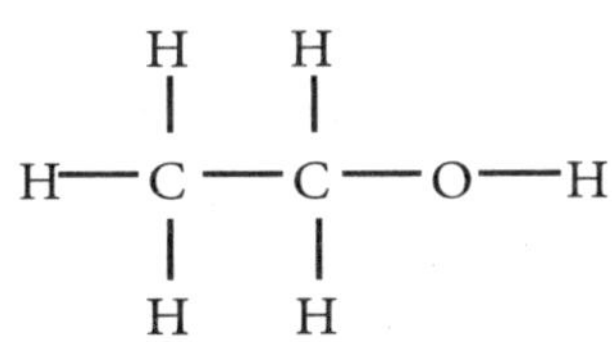

Ethanol

$$\text{The total bond dissociation energy of ethanol} = 5 \times \text{C–H} + \text{C–C} + \text{C–O} + \text{O–H}$$
$$= 5 \times 413 + 347 + 358 + 464 = 3234 \text{ kJ}$$

4.8 Calculation of enthalpy of a reaction using bond enthalpies

The enthalpy change of a reaction can be calculated using bond enthalpies. It can be assumed that all the bonds in the reactants are broken and all the bonds in the products are formed. The enthalpy change can be calculated from the bond energies using the following relation,

$$\Delta H_{(rxn)} = \Sigma \, \mathbf{BDE}_{(reactants)} - \Sigma \, \mathbf{BDE}_{(products)}$$

Example 4.8 Calculation of enthalpy of reaction using bond enthalpies

Calculate the enthalpy change of the following reaction using the bond dissociation energies.

$$\underset{\text{Ethene}}{C_2H_4\,(g)} + H_2(g) \longrightarrow \underset{\text{Ethane}}{C_2H_6\,(g)}$$

Solution

We need to know all the bonds in the reactants and products, which are depicted diagrammatically here.

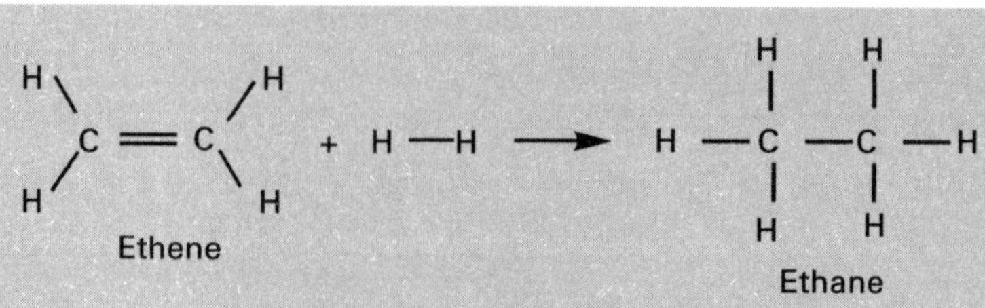

$$\Delta H_{(rxn)} = \Sigma \, BDE_{(reactants)} - \Sigma \, BDE_{(products)}$$
$$= (4 \times C\text{–}H + C\text{=}C + H\text{–}H) - (6 \times C\text{–}H + C\text{–}C)$$
$$= (4 \times 413 + 612 + 435.9) - (6 \times 413 + 347)$$
$$= 2699.9 - 2825.0 = -125.1 \text{ kJ}$$

$$C_2H_4(g) + H_2(g) \longrightarrow C_2H_6(g), \; \Delta H = -125.1 \text{ kJ}$$

Note that energy is needed to break bonds (endothermic) and energy is evolved during bond formation (exothermic). In the example given above, 2699.9 kJ heat is needed to break all the bonds in the reactants while 2825.0 kJ heat is produced when bonds are formed. Since more heat is produced than used up, the reaction is exothermic and the net heat change is negative.

Tutorial: helping you learn

Progress questions

4.1 In an experiment it was found that, 3.05 kJ of heat was evolved when 10.0 g of potassium fluoride solid was dissolved in excess of water. Calculate the heat of solution of potassium fluoride. (A_r : F = 19, K = 39.1)

4.2 The heat of combustion of methane, ΔH_c, is –890.3 kJ mol^{-1}. Calculate the heat evolved during the combustion of a kilogram of methane. (A_r : H = 1, C = 12)

4.3 Graphite is the most stable allotrope of carbon and ΔH°_f (C, graphite) = 0 kJ mol^{-1}. Given the following enthalpy changes, calculate the heat of formation, ΔH°_f (C, diamond) from C, graphite.

$$\text{C (graphite)} + O_2(g) \longrightarrow CO_2(g) \quad \Delta H_c = -393.5 \text{ kJ mol}^{-1}$$
$$\text{C (diamond)} + O_2(g) \longrightarrow CO_2(g) \quad \Delta H_c = -395.4 \text{ kJ mol}^{-1}$$

4.4 Draw an enthalpy diagram for the following changes and indicate the enthalpy of vaporisation of water on your diagram.

$$H_2(g) + \frac{1}{2}O_2(g) \longrightarrow H_2O(g), \quad \Delta H = -245.1 \text{ kJ mol}^{-1}$$
$$H_2(g) + \frac{1}{2}O_2(g) \longrightarrow H_2O(l), \quad \Delta H = -285.8 \text{ kJ mol}^{-1}$$

4.5 Calculate the enthalpy of reaction between methane and chlorine, using the bond enthalpies given in Table 4.1.

$$CH_4(g) + 4Cl_2(g) \longrightarrow CCl_4(g) + 4HCl(g)$$

4.6 Calculate the enthalpy of combustion of ethanol, using the given enthalpies of formation.

$$C_2H_5OH(l) + 3O_2(g) \longrightarrow 2CO_2(g) + 3H_2O(l)$$

Substance	ΔH°_f
C(s)	0
$CO_2(g)$	–393.5
$H_2O(l)$	–285.8
$C_2H_5OH(l)$	–277.7

4.7 Calculate the enthalpy of reaction,

$$4NH_3(g) + 5O_2(g) \longrightarrow 4NO(g) + 6H_2O(l)$$

(ΔH°_f (kJ mol^{-1}): $NH_3(g)$ = –41.1, $O_2(g)$ = 0, NO(g) = +90.3, $H_2O(l)$ = –285.8)

4.8 Calculate the enthalpy of formation of benzene, using the enthalpy of combustion of benzene and the enthalpies of formation of carbon dioxide and water.

$$C_6H_6(l) + 7\frac{1}{2}O_2(g) \longrightarrow 6CO_2(g) + 3H_2O(l),\ \ \Delta H_c = -\ 3267.4\ \text{kJ mol}^{-}$$

$$(\Delta H_f^\circ \text{ in kJ mol}^{-1}: O_2(g) = 0, CO_2(g) = -393.5, H_2O(l) = -285.8)$$

4.9 Calculate the enthalpy of the reaction shown below using the bond enthalpies given in Table 4.1

$$N_2(g) + 3H_2(g) \longrightarrow 2NH_3(g)$$

5 Equilibria

One-minute summary

Many physical and chemical changes can take place to completion under suitable conditions. In this chapter we shall examine reactions that do not go to completion, i.e., systems which contain reactants and products at equilibrium, the dynamic nature of such systems, and the industrial applications of such studies. We shall look at *equilibrium constant* and calculations involving it.

Some of the important terms in this chapter are:

- dynamic nature of equilibrium
- equilibrium constant, K_c
- equilibrium constant, K_p
- homogeneous and heterogeneous reactions
- the effect of changing pressure, concentration and temperature of systems

5.1 The dynamic nature of equilibrium

To understand the dynamic nature of equilibrium, let us consider the reaction between hydrogen and iodine. When hydrogen and iodine are heated together, hydrogen iodide is formed. On the other hand, if hydrogen iodide is heated, it dissociates into hydrogen and iodine. Such a reaction is called a **reversible reaction.** A reversible reaction is written using a double-headed arrow (as shown below) to show that the reaction can take place in either direction.

$$H_2(g) + I_2(g) \rightleftharpoons 2HI(g)$$

When hydrogen and iodine are heated in a closed container at a constant temperature (say 500 K), some hydrogen and iodine molecules start to react to form hydrogen iodide. When a few hydrogen iodide molecules are formed, some of these molecules start to dissociate to form hydrogen and iodine.

The rate of a reaction depends on the concentrations of the reactants. Initially, the rate of the forward reaction is high as the concentrations of hydrogen and iodine are high and the rate of the backward reaction is zero as there is no hydrogen iodide. As time goes on, the concentrations of hydrogen and iodine decrease and the rate of the forward reaction decreases; and the rate of the reverse reaction increases as more and more hydrogen iodide is formed. There comes a time when the rate of the forward reaction equals the rate of the reverse reaction, and the system is then said to be at ***equilibrium***.

An equilibrium system is ***dynamic*** in nature. This means that forward and reverse reactions take place simultaneously and since they occur at the same speed at equilibrium, there is no change in the concentrations of the substances. Note that though the concentrations of the substances remain constant at equilibrium at constant temperature, this does not mean that they are equal. The concentrations depend on the quantities of the substances started with.

Properties of an equilibrium system

1. Equilibrium is achieved only in a 'closed system' where substances are not allowed to escape and no new materials are added.
2. Macroscopic properties such as concentrations of substances are constant at equilibrium.
3. An equilibrium system is dynamic in nature, that is, the reaction takes place in both directions.
4. At equilibrium, the rate of the forward reaction equals the rate of the backward reaction at constant temperature.

5.2 Equilibrium constant, K_c

If an equilibrium mixture contains the reactants A and B and the products C and D, related by the stoichiometric equation

$$aA + bB \rightleftharpoons cC + dD$$

the equilibrium constant K_c is expressed by,

$$K_c = \frac{[C]^c[D]^d}{[A]^a[B]^b}$$

For a given equilibrium reaction, *K_c is a constant at constant temperature.* The square brackets represent the equilibrium concentrations in mol dm^{-3} and c (in K_c) stands for concentration. The unit for K_c is derived from the concentration terms in the expression for K_c and it depends on the coefficients *a*, *b*, *c* and *d*. Examples of writing expressions for K_c and deriving its unit are given below in Examples 5.1 and 5.2.

Example 5.1 Write an expression for K_c for the following reaction and work out its unit.

$$H_2(g) + I_2(g) \rightleftharpoons 2HI(g)$$

Solution

$$K_c = \frac{[HI]^2}{[H_2][I_2]} \qquad \frac{(\cancel{mol\ dm^{-3}})^2}{\cancel{mol\ dm^{-3}} \cdot \cancel{mol\ dm^{-3}}}$$

K_c has no unit.

Example 5.2 Write an expression for K_c for the dissociation of ammonia into hydrogen and nitrogen and work out its unit.

$$2NH_3(g) \rightleftharpoons N_2(g) + 3H_2(g)$$

Solution

$$K_c = \frac{[N_2][H_2]^3}{[NH_3]^2} \qquad \frac{mol\ dm^{-3} \cdot (mol\ dm^{-3})^{\cancel{3}}}{(\cancel{mol\ dm^{-3}})^2} = mol^2\ dm^{-6}$$

Note that two concentration terms are cancelled out from the numerator and denominator and two concentration terms remain in the numerator.

Example 5.3 Write an expression for K_c for the following reaction.

$$\underset{\text{Ethanoic acid}}{CH_3COOH(l)} + \underset{\text{Ethanol}}{C_2H_5OH(l)} \rightleftharpoons \underset{\text{Ethyl ethanoate}}{CH_3COOC_2H_5(l)} + H_2O(l)$$

Solution

$$K_c = \frac{[CH_3COOC_2H_5][H_2O]}{[CH_3COOH][C_2H_5OH]}$$

Example 5.4 Calculation of equilibrium constant, K_c

Consider the equilibrium reaction

$$CO(g) + Cl_2(g) \rightleftharpoons COCl_2(g)$$

Given the concentrations at equilibrium at a particular temperature; [CO] = 0.008 mol dm^{-3}, $[Cl_2]$ = 0.065 mol dm^{-3} and $[COCl_2]$ = 0.11 mol dm^{-3}, calculate the equilibrium constant K_c for the formation of $COCl_2$. Work out the unit for K_c.

Solution

$$K_c = \frac{[COCl_2]}{[CO][Cl_2]}$$

$$= \frac{0.11 \cancel{\text{mol dm}^{-3}}}{0.008 \cancel{\text{mol dm}^{-3}} \times 0.065 \text{ mol dm}^{-3}}$$

$$= 211.5 \text{ mol}^{-1} \text{ dm}^3$$

Example 5.5

1. Write an expression for the equilibrium constant for the reaction

$$2HI(g) \rightleftharpoons H_2(g) + I_2(g)$$

2. How is the equilibrium constant for the forward reaction related to that for the reverse reaction?

Solution

1. Equilibrium constant for the forward reaction, K_c, is

$$K_c = \frac{[H_2][I_2]}{[HI]^2}$$

2. Equilibrium constant for the reverse reaction, K'_c, is

$$K'_c = \frac{[HI]^2}{[H_2][I_2]}$$

K_c is the reciprocal of K'_c.

$$K'_c = \frac{1}{K_c}$$

Homogeneous and heterogeneous reactions

Reactions in which all the reactants and products are in the same phase are called **homogeneous reactions**. The hydrogen-iodine-hydrogen iodide system consists of gases. All substances in the esterification reaction in which ethanoic acid reacts with ethanol to form ethyl ethanoate and water (Example 5.3) are in the liquid state. These are homogeneous reactions.

Consider the decomposition of calcium carbonate to calcium oxide and carbon dioxide. This is a **heterogeneous reaction** since the substances in the system are in different phases. Calcium carbonate and calcium oxide are solids and carbon dioxide a gas.

$$CaCO_3(s) \rightleftharpoons CaO(s) + CO_2(g)$$

The equilibrium constant K_c for this reaction is given by the expression

$$K_c = [CO_2] \text{ mol dm}^{-3}$$

The equilibrium concentrations of solids are taken as constant and do not appear in the expression for the equilibrium constant.

> **Practice question 5.1** Write an expression for K_c and determine its unit for each of the following equilibrium systems.
>
> (1) $COCl_2(g) \rightleftharpoons CO(g) + Cl_2(g)$
>
> (2) $2PbS(s) + 3O_2(g) \rightleftharpoons 2PbO(s) + 2SO_2(g)$

5.3 Equilibrium constant, K_p

For a gaseous reaction at equilibrium, the equilibrium constant is denoted by K_p when the amounts of the substances in the system are expressed in partial pressures. Consider the reaction between nitrogen and hydrogen to form ammonia.

$$N_2(g) + 3H_2(g) \rightleftharpoons 2NH_3(g)$$

The equilibrium constant K_p is expressed by,

$$K_p = \frac{P^2_{NH_3}}{P_{N_2} \times P^3_{H_2}}$$

Like K_c, K_p is a constant for a given reaction at constant temperature. P in the expression stands for the partial pressure of a gas. The unit for K_p is derived individually as for K_c.

> **Notes**
>
> 1. The partial pressure of a gas is the pressure exerted by that gas if it alone were present in a container.
> 2. The total pressure of a mixture of gases is equal to the sum of the partial pressures of all the gases.

3. The unit for pressure is atmosphere (atm.) or Pascal (Pa) or kilo Pascal (kPa).
   1 atm = 1.01×10^5 Pa or 101 kPa.

Example 5.6 Derive the unit for K_p for the reaction.

$$N_2(g) + 3H_2(g) \rightleftharpoons 2NH_3(g)$$

Solution

$$K_p = \frac{P^2_{NH_3}}{P_{N_2}\ P^3_{H_2}}$$

$$\text{Units for } K_p = \frac{atm^2}{atm \times atm^3} = atm^{-2}$$

Example 5.7 Calculation of equilibrium constant, K_p

For the gaseous reaction

$$PCl_5(g) \rightleftharpoons PCl_3(g) + Cl_2(g)$$

the equilibrium partial pressures of PCl_5, PCl_3 and Cl_2 are 0.01 atm, 0.4 atm and 0.62 atm, respectively. Work out the value of K_p together with its unit.

Solution

$$\begin{aligned} K_p &= \frac{P_{PCl_3} \times P_{Cl_2}}{P_{PCl_5}} \\ &= \frac{0.4\ atm \times 0.62\ atm}{0.01\ atm} \\ &= 24.8\ atm \end{aligned}$$

Practice question 5.2 Write expressions for K_p for each of the following reactions and work out the units if the partial pressures are in atmospheres.

(1) $2SO_2\ (g) + O_2(g) \rightleftharpoons 2SO_3(g)$
(2) $COCl_2(g) \rightleftharpoons CO(g) + Cl_2(g)$

5.4 Le Chatelier's principle

Le Chatelier's principle states that if a stress (such as changing concentration, pressure or temperature) is applied to a system at equilibrium, the system adjusts in a direction so as to cancel the effect of the change.

The importance of Le Chatelier's principle is that it can be used to predict the direction of a change or reaction when the pressure, concentration or temperature of the system at equilibrium is changed.

5.5 Effect of a change in concentration

Consider the nitrogen-hydrogen-ammonia system at equilibrium at a constant temperature.

$$N_2(g) + 3H_2(g) \rightleftharpoons 2NH_3(g)$$

If the concentration of one of the reactants is increased, the system tries to compensate for this by reducing the concentrations of the reactants by moving the equilibrium position to the right. Imagine that some hydrogen is added to the system at equilibrium. According to Le Chatelier's principle, to minimise the effect of the change, the hydrogen molecules will react with nitrogen to form ammonia. This will proceed to the extent that the ratio for the equilibrium constant is kept the same. Similarly if some ammonia is added to the reaction mixture, the equilibrium moves to the left, this results in an increase in the amounts of nitrogen and hydrogen present. If some ammonia is removed from the system the forward reaction takes place. **If the concentration of a reactant is increased or the concentration of a product is decreased, the equilibrium will shift towards the products, i.e., more products are formed. If the concentration of a reactant is decreased or the concentration of a product is increased the equilibrium will shift towards the reactants, i.e., more reactants are formed.**

5.6 Effect of changing pressure

Changes in pressure will have very little effect on systems containing solids, liquids and solutions. If however the pressure in a gaseous system is increased, the system will try to lower the pressure by lowering the number of molecules, since pressure is proportional to the number of molecules. Consider again, the nitrogen-hydrogen-ammonia system.

$$\underset{4\,\text{molecules}}{N_2(g) + 3H_2(g)} \rightleftharpoons \underset{2\,\text{molecules}}{2NH_3(g)}$$

Four molecules of the reactants react to give two molecules of the products. If the total pressure of this system is increased, the system tends to reduce the pressure by favouring the forward reaction. This is because if the forward reaction predominates the number of molecules will be reduced and so, pressure lowered. So, increasing the pressure favours the production of ammonia. Similarly if the pressure is lowered, ammonia dissociates to produce more molecules (of N_2 and H_2) in order to increase the pressure. In other words, if the number of reactant molecules is higher than the number of product molecules, increasing pressure favours the forward reaction. Similarly, lowering pressure favours the backward reaction. **An increase in pressure will shift the equilibrium towards the side having fewer gaseous molecules.**

A system such as hydrogen-iodine-hydrogen iodide in which the number of the reactant molecules equals the number of the product molecules *will not be affected* by a change in pressure.

$$\underset{2\,\text{molecules}}{H_2(g) + I_2(g)} \rightleftharpoons \underset{2\,\text{molecules}}{2HI(g)}$$

5.7 Effect of changing temperature

The equilibrium constant for a reaction is constant at constant temperature and is altered by changing the temperature. **If the temperature is changed the value of K will also be changed.** K will increase or decrease depending on whether the reaction is exothermic or endothermic.

The formation of ammonia from nitrogen and hydrogen is an exothermic reaction.

$$N_2(g)+3H_2(g) \rightleftharpoons 2NH_3(g),\ \Delta H=-92.0\,kJ$$

Consider a system containing hydrogen, nitrogen and ammonia at equilibrium at a constant temperature. If the temperature of the mixture is raised, the system will try to lower the temperature by favouring the direction of the endothermic reaction which absorbs heat. The decomposition of ammonia is endothermic. So by raising the temperature ammonia decomposes, the concentration of ammonia is lowered and the concentrations of nitrogen and hydrogen are increased. **An increase in temperature favours an endothermic reaction and a lowering of temperature favours an exothermic reaction.**

Now, let us examine how the change in temperature affects the value of the equilibrium constant, K. For the above reaction at a higher temperature, the concentration of the product (ammonia) is lower than its concentration at a lower temperature. If the concentration of the product is lower, the ratio of the [product] to the [reactant] will be lower. Hence, the value of K is lower at a higher temperature. **As the temperature is increased, the value of K decreases for an exothermic reaction and increases for an endothermic reaction.**

5.8 Effect of adding a catalyst

A catalyst is a substance that increases the speed of a reaction without being consumed itself (Section 6.3). A catalyst increases the speed of both forward and backward reactions and equilibrium is reached in a shorter time. **A catalyst has no effect on a system at equilibrium.**

Applications

Le Chatelier's principle helps us to predict the direction of a reaction when changes in temperature, pressure or concentration are made on an equilibrium system. Many reactions, including many of the industrial processes, are reversible. It is important to employ suitable conditions for the maximum yield of the product/s. These suitable conditions are predicted by applying Le Chatelier's principle.

We have discussed the effect of changing pressure, temperature and concentration on the N_2—H_2—NH_3 system in the previous Sections. Let us consider another example; an esterification reaction. The reaction between ethanoic acid and ethanol to form ethyl ethanoate and water is a reversible reaction.

$$\underset{\text{Ethanoic acid}}{CH_3COOH(l)} + \underset{\text{Ethanol}}{C_2H_5OH(l)} \rightleftharpoons \underset{\text{Ethyl ethanoate}}{CH_3COOC_2H_5(l)} + H_2O(l)$$

If the reaction is carried out starting with 1 mole each of the reactants at room temperature, an equilibrium system is formed when about the two-third of the reactants react to give the products. About one-third of the reactants remain in the equilibrium system. One way of moving the reaction more to the forward direction is by removing water formed during the reaction. This can be done by adding a dehydrating agent like concentrated sulphuric acid to the reaction mixture.

5.9 Equilibria and industrial processes

The Haber process

Ammonia is an important chemical which is used in the preparation of nitric acid, fertilisers and explosives. Ammonia is prepared industrially from hydrogen and nitrogen by a process called the **Haber process.** Hydrogen and nitrogen are mixed in the ratio 3:1 and allowed to react under suitable conditions. The formation of ammonia is exothermic and it is a reversible reaction.

What are the best conditions for the maximum yield of ammonia?

$$\underset{1\text{ mol}}{N_2(g)} + \underset{3\text{ mol}}{3H_2(g)} \rightleftharpoons \underset{2\text{ mol}}{2NH_3(g)}, \quad \Delta H = -92.0 \text{ kJ}$$

1. **Pressure** As explained above, in Section 5.6, a rise in pressure favours the production of ammonia as four molecules of the reactants react to form two molecules of the product. At high pressures, ammonia gas is liquefied and removed from the gaseous reaction mixture. This again favours the forward reaction.
2. **Temperature** Since the formation of ammonia is an exothermic reaction, low temperatures favour the forward reaction (Section 5.7). But at low temperatures the rate of the reaction is low and not favourable economically. At high temperatures the rate of the reaction is high, but backward reaction or the dissociation of ammonia takes place. So a temperature which is neither too low nor too high is employed.
3. **Catalyst** A catalyst is used to increase the speed of the reaction.

Favourable conditions for the production of ammonia are

- a pressure of about 150 atm.
- a temperature between 670 to 770 K.
- a catalyst of porous iron containing metal oxides which act as promoters.

The ammonia produced is continuously removed from the system as liquid ammonia, and the unreacted gases are recycled.

The Contact process

Sulphuric acid is another important chemical and it is used in the production of fertilizers, dye-stuffs, paints, petrochemicals and detergents. Sulphuric acid is prepared commercially by the **contact process.** Sulphur is the starting material, it is burned in oxygen to form sulphur dioxide.

$$S(s) + O_2(g) \longrightarrow SO_2(g)$$

Sulphur dioxide is oxidised to sulphur trioxide by heating with oxygen in the presence of vanadium (V) oxide catalyst.

$$2SO_2(g) + O_2(g) \rightleftharpoons 2SO_3(g), \ \Delta H = -196 \text{ kJ}$$

Sulphur trioxide is dissolved in concentrated sulphuric acid to produce oleum, which is mixed with water to give sulphuric acid.

$$SO_3(g) + H_2SO_4(l) \longrightarrow \underset{\text{Oleum}}{H_2S_2O_7(l)}$$

$$H_2S_2O_7(l) + H_2O(l) \longrightarrow 2H_2SO_4(l)$$

The formation of sulphur trioxide from sulphur dioxide and oxygen is a reversible reaction and for the maximum conversion of the reactants to the products, optimum conditions should be employed.

Pressure According to Le Chatelier's principle, high pressures favour the production of sulphur trioxide as three molecules of the reactants give two molecules of the products.

Concentration of reactants According to Le Chatelier's principle, if the concentration of one of the reactants is increased, the forward reaction is favoured. Oxygen is a cheap raw material since it can be obtained from air. Using excess oxygen ensures a high conversion of SO_2 to SO_3.

Temperature Since the formation of sulphur trioxide is an exothermic reaction a low temperature favours the production of sulphur trioxide. As we have seen earlier, at low temperatures, the reaction rate is low and to speed up the reaction a temperature of about 720 K is used.

Catalyst To increase the speed of the reaction vanadium (V) oxide (V_2O_5) catalyst is used.

Example 5.8 0.34 mole of ethanoic acid and 0.32 mole of ethanol were placed in a container and allowed to reach equilibrium at 291 K. The equilibrium mixture contained 0.12 mole of unreacted ethanoic acid.

1. Write a balanced equation for the reaction.
2. Calculate the value of the equilibrium constant K_c at 291 K.
3. Work out the unit for K_c.

Solution

(1) $$\underset{\text{Ethanoic acid}}{CH_3COOH(l)} + \underset{\text{Ethanol}}{C_2H_5OH(l)} \rightleftharpoons \underset{\text{Ethyl ethanoate}}{CH_3COOC_2H_5(l)} + H_2O(l)$$

(2) **Hint:** 1. Write the expression for K_c for the reaction. 2. Calculate the equilibrium concentrations of all the substances from the given data as shown below. 3. The number of moles of ethanoic acid reacted is obtained by substracting the amount of it remaining from the amount taken.

$$K_c = \frac{[CH_3COOC_2H_5][H_2O]}{[CH_3COOH][C_2H_5OH]}$$

	CH_3COOH (l) +	C_2H_5OH (l)	$\rightleftharpoons$	$CH_3COOC_2H_5$ (l) +	H_2O (l)
Taken:	0.34 mol	0.32 mol		-	-
Remain:	0.12 mol				
Reacted:	0.22 mol	0.22 mol		-	-
At equilibrium:	0.12 mol	0.10 mol		0.22 mol	0.22 mol

$$K_c = \frac{[CH_3COOC_2H_5][H_2O]}{[CH_3COOH][C_2H_5OH]}$$

$$= \frac{0.22 \cancel{mol} \times 0.22 \cancel{mol}}{0.12 \cancel{mol} \times 0.10 \cancel{mol}}$$

$$= 4.03$$

(3) K_c has no unit for this reaction.

Tutorial: helping you learn

Practice questions

5.3 Calculate the equilibrium constant K_p for the reaction

$$2NO_2(g) \rightleftharpoons 2NO(g) + O_2(g)$$

given the equilibrium partial pressures at a particular temperature:

$P_{NO_2} = 0.2$ atm, $P_{NO} = 0.2$ atm and $P_{O_2} = 150$ atm

5.4 The formation of nitric oxide gas is an endothermic reaction.

$$N_2(g) + O_2(g) \rightleftharpoons 2NO(g), \ \Delta H = +90 \text{ kJ mol}^{-1}$$

How would you expect K_c to change with temperature? Explain.

Progress questions

5.1 Write an expression for the equilibrium constant K_c for each of the following reactions, working out its unit in each case.

$$2SO_3(g) \rightleftharpoons 2SO_2(g) + O_2(g)$$
$$PCl_3(g) + Cl_2(g) \rightleftharpoons PCl_5(g)$$
$$N_2(g) + O_2(g) \rightleftharpoons 2NO(g)$$
$$CH_3COOCH_3(l) + H_2O(l) \rightleftharpoons CH_3COOH(l) + CH_3OH(l)$$

5.2 Write an expression for the equilibrium constant K_p for each of the following reactions.

$$2SO_3(g) \rightleftharpoons 2SO_2(g) + O_2(g)$$
$$PCl_3(g) + Cl_2(g) \rightleftharpoons PCl_5(g)$$
$$N_2(g) + O_2(g) \rightleftharpoons 2NO(g)$$

5.3 The dissociation of phosphorus pentachloride to phosphorus trichloride and chlorine is endothermic. Consider the equilibrium system at 400 K.

$$PCl_5(g) \rightleftharpoons PCl_3(g) + Cl_2(g)$$

Describe, giving reasons, the direction of the net reaction with each of the following changes.

(a) increasing the total pressure of the system
(b) raising the temperature at constant pressure
(c) adding more chlorine.

5.4 For the reaction

$$2SO_2(g) + O_2(g) \rightleftharpoons 2SO_3(g), \ \Delta H = -198 \text{ kJ}$$

predict the effect on the equilibrium when the following changes are made

(a) adding more oxygen
(b) decreasing the pressure at constant temperature
(c) raising the temperature at constant pressure
(d) adding a catalyst.

5.5 The equilibrium constant K_c for the reaction,

$$H_2(g) + I_2(g) \rightleftharpoons 2HI(g)$$

is 2916 at 700 K. Use this information to calculate the equilibrium constant for the formation of hydrogen iodide.

$$\tfrac{1}{2}H_2(g) + \tfrac{1}{2}I_2(g) \rightleftharpoons HI(g)$$

6 Chemical Kinetics

One-minute summary

Chemical kinetics is the study of reaction rates. The rate of a reaction is the amount of a substance reacted or a product formed in unit time. The rate of a reaction depends on a number of factors such as the nature and concentrations of the reactants, temperature, surface area of a solid reactant and presence of a catalyst. Collision theory states that the rate of a reaction is proportional to the number of effective collisions between reactant molecules. We shall discuss the factors which affect reaction rates in terms of collision theory.

We shall examine in this chapter:

- the rate of a reaction
- collision theory
- energy level diagrams
- Maxwell-Boltzmann distribution of energies
- activation energy
- factors which affect reaction rates
- heterogeneous catalysts
- homogeneous catalysts

6.1 Rates of reactions: introduction

The study of rates of reactions is called **chemical kinetics.** You will recall that different reactions take place at different speeds. The neutralisation reaction between hydrochloric acid and sodium hydroxide is instantaneous. When a piece of magnesium ribbon is placed in aqueous hydrochloric acid, bubbles of hydrogen gas are produced and the magnesium dissolves and disappears gradually. This is a slower reaction. Iron rusts in air even more slowly, taking weeks.

The **rate of a reaction** is the amount of a substance reacted or formed in unit time, usually in one second.

For the reaction,

$$2HI(g) \longrightarrow H_2(g) + I_2(g)$$

the rate of dissociation of hydrogen iodide is 5.0×10^{-3} mol dm^{-3} s^{-1}, at a certain temperature. This expresses the amount of hydrogen iodide in mol dm^{-3} dissociated per second at that temperature. The rate of formation of hydrogen or iodine is 2.5×10^{-3} mol dm^{-3} s^{-1}, which is half of the rate of dissociation of hydrogen iodide, as *two* molecules of hydrogen iodide react to give *one* molecule each of hydrogen and iodine.

6.2 Collision theory

Consider a gaseous reaction, say, the dissociation of hydrogen iodide. A gaseous reactant in a container consists of tiny molecules moving in all directions at a certain average speed. The average speed of the molecules depends on their average kinetic energy which in turn is proportional to absolute temperature. The molecules collide with one another and these molecular collisions lead to the formation of the product.

Collision theory states that the rate of a reaction is proportional to the number of collisions per second. At a higher temperature, molecules have greater kinetic energy, they move with greater speed and the number of collisions per second is higher. This is one of the reasons for a higher reaction rate at a higher temperature. A ten degree rise in temperature from room temperature doubles the reaction rate of many reactions.

If all molecular collisions lead to formation of products, all gaseous reactions must be instantaneous, as the number of collisions per second in a gaseous system is very high. But that is not the case. According to the collision theory, the rate of a reaction is proportional to the number of *effective collisions* per second. What is an effective collision? To understand this, let us discuss (1) the activation energy and (2) the Maxwell-Boltzmann distribution of energies in a sample of gas molecules.

The Maxwell-Boltzmann distribution of energies

At a certain temperature (T_1), the molecules of a gas possess a certain average value of kinetic energy (K_1) (Figure 6.1). This does not mean that all the molecules have the same average value of kinetic energy; some have lower than the average value, and some have higher. Energy can be transferred between molecules during molecular collisions. The Maxwell-Boltzmann distribution of energies (Figure 6.1) shows the fraction of molecules that have certain energy values at a temperature, T_1.

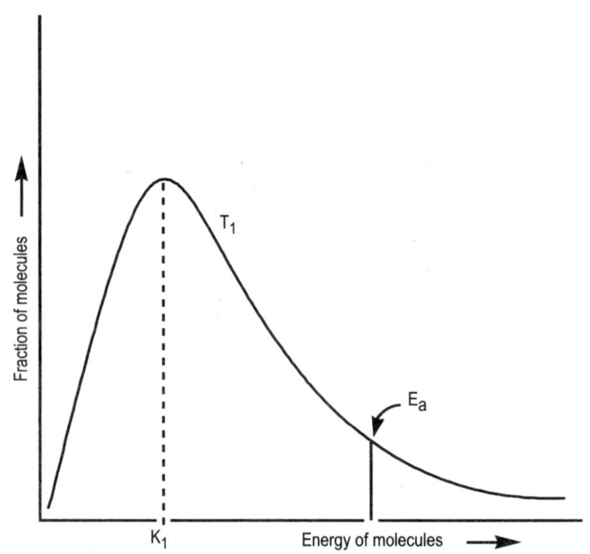

Figure 6.1: Maxwell-Boltzmann distribution of energies

Let us look at the 'distribution of energies' curves at two different temperatures, T_1 and T_2 where T_2 is higher than T_1 (Figure 6.2). The average kinetic energy of molecules (K_2) is higher at a higher temperature (T_2). At low temperatures there are more molecules which have energies close to the average value. At higher temperatures, there is a more even distribution of energies, and a smaller fraction of molecules have energies close to the average value.

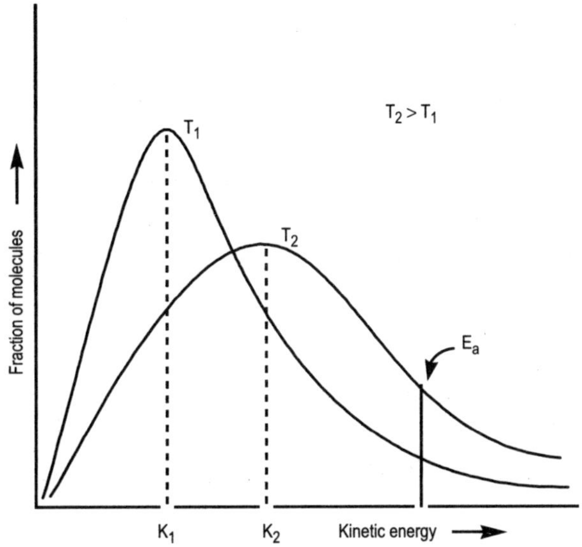

Figure 6.2: Maxwell-Boltzmann distribution of energies at two temperatures

Activation energy

In a gas, molecules move in all possible directions. Consider a gaseous reaction. When the reactant molecules approach each other their outer electrons repel and molecules may move apart. If the molecules possess enough energy to overcome the repulsive forces, they collide effectively which will lead to the formation of an **intermediate complex.** The intermediate complex is also called the **activated complex.** This complex decomposes to give the products or changes back to the reactants (Figure 6.3). The minimum amount of energy needed for the formation of the activated complex is called the **activation energy (E_a).**

At a lower temperature, fewer molecules posses activation energy causing a lower rate of reaction. At a higher temperature, there are more molecules which possess energies equal to or greater than the activation energy, leading to a higher rate of reaction. Note that the fraction of molecules which possess activation energy at a lower temperature is represented by the area below the curve for T_1, and at a higher temperature below the curve for T_2 (Figure 6.2).

The activation energy for a reaction can be depicted diagrammatically as shown below (Figure 6.3). This is called an energy level diagram where the energy of the reactants, the intermediate and the products are plotted against the reaction co-ordinate.

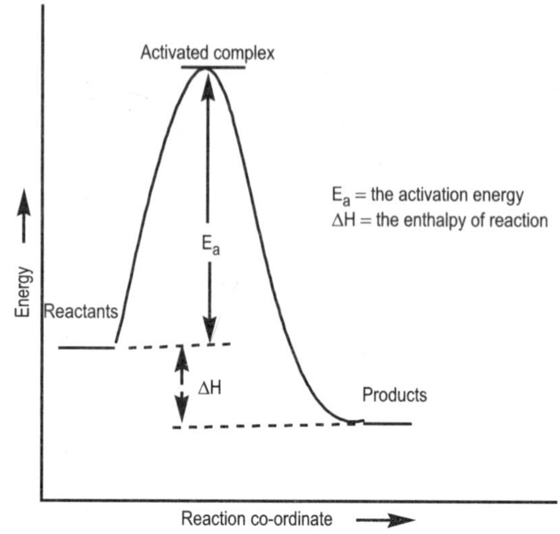

Figure 6.3: Energy level diagram for an exothermic reaction

Note that the energy level of the products is lower than that of the reactants for an exothermic reaction, and activation energy is the amount of energy needed by the reactants for the formation of the intermediate.

6.3 Factors affecting reaction rates

1. Nature of the reactants

Some substances are more reactive than others. Some reactions are faster than others. Reaction rate depends on the nature of the reactants. Let us consider the reaction between sodium hydroxide and hydrochloric acid; this is a fast reaction. The reason is that aqueous sodium hydroxide and hydrochloric acid are fully ionised in solution, and neutralisation involves the union of positive hydrogen ions from the acid and negative hydroxide ions from the base, to form water. This reaction takes place as fast as the solutions are mixed.

$$NaOH(aq) + HCl(aq) \longrightarrow NaCl(aq) + H_2O(l)$$
$$OH^-(aq) + H^+(aq) \longrightarrow H_2O(l)$$

As another example, consider the reaction between sodium hydroxide and ethyl ethanoate to form ethanol and sodium ethanoate. The reaction is slow since there are covalent bonds to be broken in ethyl ethanoate molecules before reaction can take place.

$$\underset{\text{Ethyl ethanoate}}{CH_3COOC_2H_5(l)} + NaOH(aq) \longrightarrow \underset{\text{Ethanol}}{C_2H_5OH(aq)} + \underset{\text{Sodium ethanoate}}{CH_3COONa(aq)}$$

$$CH_3COOC_2H_5(l) + OH^-(aq) \longrightarrow C_2H_5OH(aq) + CH_3COO^-(aq)$$

Reactions which require low activation energies proceed faster than those which require high activation energies.

2. Concentration of the reactants (or pressure for gaseous reactions)

Increasing the concentration of a reactant generally increases the rate of reaction. Consider the reaction between magnesium and hydrochloric acid. A piece of magnesium ribbon dissolves faster in 1 mol dm^{-3} hydrochloric acid than in 0.1 mol dm^{-3} hydrochloric acid.

$$Mg(s) + 2HCl(aq) \longrightarrow MgCl_2(aq) + H_2(g)$$

Increasing the concentration of hydrochloric acid increases the number of collisions per second thus increasing the rate of reaction. Similarly if the partial pressure of a gaseous reactant is increased, its concentration is increased and so is the reaction rate.

3. Physical state of the reactants

This factor is important in reactions involving solids with liquids or solutions. The rate of reaction between calcium carbonate solid in the form of lumps and hydrochloric acid is slower than that between calcium carbonate powder and hydrochloric acid of the same concentration. This is because surface area is increased when lumps are powdered, and so number of collisions increases as does reaction rate.

$$CaCO_3(s) + 2HCl(aq) \longrightarrow CaCl_2(aq) + CO_2(g) + H_2O(l)$$

4. Catalyst

A **catalyst** is a substance that increases the rate of a reaction without itself being used up. The rates of many reactions are increased by adding suitable catalysts. Take the thermal decomposition of potassium chlorate to potassium chloride and oxygen as an example. It is a slow reaction which can be speeded up by adding a small quantity of manganese (IV) oxide (MnO_2) as a catalyst.

$$2KClO_3(s) \xrightarrow{MnO_2} 2KCl(s) + 3O_2(g)$$

Vanadium (V) oxide (V_2O_5) is used as a catalyst for the conversion of sulphur dioxide to sulphur trioxide, and sulphuric acid is used as a catalyst for the hydrolysis of an ester (say ethyl ethanoate), to form a carboxylic acid and an alcohol.

$$2SO_2(g) + O_2(g) \xrightarrow{V_2O_5} 2SO_3(g)$$

$$\underset{\text{Ethyl ethanoate}}{CH_3COOC_2H_5(l)} + H_2O(l) \xrightarrow{H_2SO_4} \underset{\text{Ethanoic acid}}{CH_3COOH(l)} + \underset{\text{Ethanol}}{C_2H_5OH(l)}$$

Studies show that catalysts take part in the reactions and are then regenerated. *The catalyst usually forms an intermediate complex through an alternative reaction pathway which requires less activation energy.* When the activation energy is lowered, there are more molecules with the required activation energy values and so the reaction rate increases. This is shown diagrammatically (Figure 6.4) for the reaction,

Note in Figure 6.4 that the enthalpy change, ΔH, for the reaction remains the same for the catalysed and uncatalysed reactions and the function of the catalyst is to lower the activation energy for the reaction by taking an alternative reaction path.

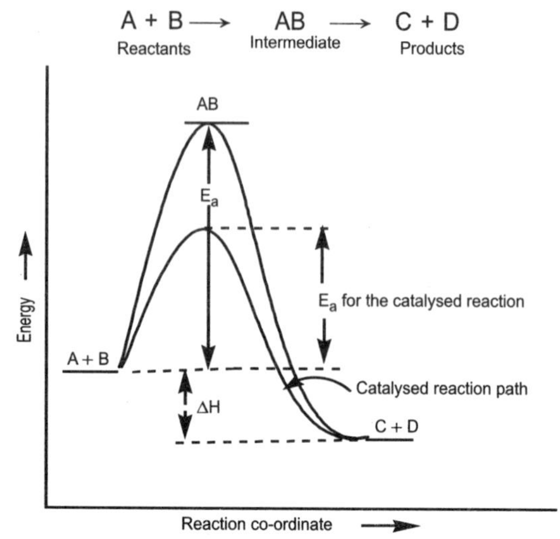

Figure 6.4: Energy level diagram for a catalysed reaction

A catalyst is often specific for a particular reaction. Iron for the manufacture of ammonia from nitrogen and hydrogen, nickel powder for the hydrogenation of oils and the enzyme zymase for the fermentation of glucose are examples of specific catalysts.

Many gaseous and liquid phase reactions use solid catalysts. A **heterogeneous catalyst** is one which is in a different phase from the reactants. Vanadium (V) oxide solid is used as a catalyst in the preparation of sulphur trioxide gas from sulphur dioxide and oxygen gases, and nickel metal in the hydrogenation of oils into fats. These are examples of heterogeneous catalysts. The reactant molecules are adsorbed on to the surface of the catalyst by weak bonds. This helps to weaken the bonds in the adsorbed molecules and thus, bonds in the reactant molecules break more easily. New bonds are formed which lead to the formation of product molecules. Heterogeneous catalysts thus increase the speed of reactions. This can be shown diagrammatically for the reaction (Figure 6.5).

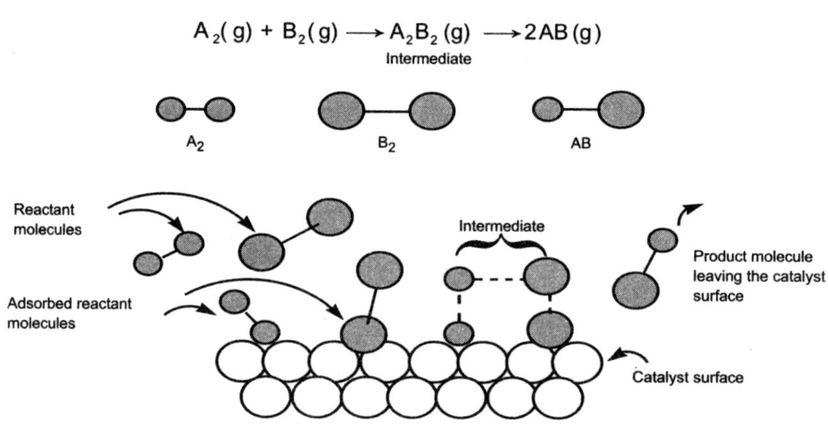

Figure 6.5

A solid catalyst in the powdered form has a greater surface area and is thus more effective than in a lump form.

A **homogeneous catalyst** is one which is in the same phase as the reactants. Aqueous sulphuric acid is a catalyst used in the hydrolysis of an ester and it is an example of a homogeneous catalyst, the catalyst and reactants all being in the liquid phase. The catalyst takes part in the reaction; in this case, the hydrogen ion of the acid catalyst reacts with the ester to form an intermediate which hydrolyses to produce a carboxylic acid and an alcohol, regenerating the hydrogen ion. Below is shown the equation for the hydrolysis of ethyl ethanoate to give ethanoic acid and ethanol.

$$\underset{\text{Ethyl ethanoate}}{CH_3COOC_2H_5\,(l)} + H_2O(l) \xrightarrow{H^+} \underset{\text{Ethanoic acid}}{CH_3COOH(l)} + \underset{\text{Ethanol}}{C_2H_5OH(l)}$$

Another example of a homogeneous catalyst is chlorine in the decomposition of ozone to oxygen. All the materials in this reaction are in the gaseous phase.

$$2O_3(g) \xrightarrow{Cl_2} 3O_2(g)$$

This reaction can take place in atmosphere. Ozone is present in small quantities in the air as are chlorine atoms from the dissociation of chlorofluorocarbons (CFCs). Cl atoms are reactive free radicals which have an unpaired electron. A Cl atom combines with an O_3 molecule to give an O_2 molecule and another free radical, ClO., which in turn reacts with O atoms formed by the dissociation of O_2 molecules by sunlight. The Cl atom regenerated reacts with more O_3 and a chain reaction takes place.

$$Cl^\bullet + O_3 \longrightarrow O_2 + ClO^\bullet$$

$$O_2 \xrightarrow{\text{Sunlight}} 2O^\bullet$$

$$ClO^\bullet + O^\bullet \longrightarrow Cl^\bullet + O_2$$

Ozone present in the stratosphere provides some protection to the earth from harmful high energy ultraviolet radiation. The depletion of the ozone layer by the above reactions is believed to be one of the reasons for the increasing incidence of skin cancer.

Tutorial: helping you learn

Progress questions

6.1 What do you understand by 'rate of a reaction'? List the factors which affect the rate of a reaction.

6.2 What is 'collision theory'? Explain the factors which affect the rate of a reaction in terms of collision theory.

6.3 Consider that a gas in a closed container consists of molecules which possessan average kinetic energy of K_1 joules at a temperature of T_1 degree Kelvin. Do all the molecules possess the same value of kinetic energy K_1? If your answer is 'No', explain briefly, with the help of an energy distribution curve, how the kinetic energy is distributed among the molecules.

6.4 What is activation energy? Give a brief explanation with the help of an energy level diagram.

6.5 Bromoethane and sodium hydroxide solution react to give ethanol and sodium bromide. The ionic equation for the reaction is

$$\underset{\text{Bromoethane}}{CH_3CH_2Br(l)} + OH^-(aq) \longrightarrow \underset{\text{Ethanol}}{CH_3CH_2OH(l)} + Br^-(aq)$$

The rate of this reaction is 1.6×10^{-4} mol dm^{-3} s^{-1} at 30° C and 5.0×10^{-4} mol dm^{-3} s^{-1} at 40° C, i.e., the reaction rate is increased almost three-fold for an increase of 10° C. Give an explanation for the increase in terms of collision theory and activation energy.

6.6 Draw an energy level diagram for the reaction

$$\underset{\text{Reactants}}{A + B} \longrightarrow \underset{\text{Intermediate}}{AB} \longrightarrow \underset{\text{Products}}{C + D}$$

with and without a catalyst. Mark clearly on the diagram

(a) the energy levels of the reactants, intermediate and products
(b) the activation energy
(c) the $\Delta H_{reaction}$.

6.7 What is a catalyst? With the help of an energy level diagram, show the function of a catalyst.

6.8 What is a homogeneous catalyst? How do you think a homogeneous catalyst alters the reaction rate?

6.9 What is a heterogeneous catalyst? Give an example of a reaction in which a heterogeneous catalyst is used and explain how the catalyst alters the rate of a reaction.

7 Redox Reactions

One-minute summary

A redox reaction is one in which oxidation and reduction take place simultaneously. In simple terms, oxidation is addition of oxygen or an electronegative element and reduction is addition of hydrogen or an electropositive element. This is one of the earliest ways of thinking about oxidation and reduction. Knowledge about the structure of atoms and chemical bonding has helped to extend the concept of oxidation and reduction in terms of electron transfer, and further in terms of change in oxidation number.

In this chapter we shall discuss:

- redox reactions in terms of electron transfer
- partial and overall ionic equations for redox reactions
- oxidation number
- redox reactions in terms of change in oxidation number
- oxidising and reducing agents

7.1 Redox reactions: in terms of electron transfer

Many chemical reactions take place between substances during which electrons are transferred from one substance to another. In terms of electron transfer, **oxidation is electron loss**, and **reduction is electron gain**.

Consider the reaction between sodium and chlorine to form sodium chloride.

$$Na(s) + \frac{1}{2}Cl_2(g) \longrightarrow NaCl(s)$$

During the reaction a sodium atom ($1s^2 2s^2 2p^6 3s^1$) gives out its valence electron and becomes a positively charged Na^+ ion. At the same time, a chlorine atom ($1s^2 2s^2 2p^6 3s^2 3p^5$) takes that electron and becomes a negatively charged Cl^- ion. These oppositely charged ions attract and give rise to the bonding in sodium chloride.

$$Na \longrightarrow Na^+ + e^-$$

$$\frac{1}{2}Cl_2 + e^- \longrightarrow Cl^-$$

$$Na^+ + Cl^- \longrightarrow NaCl$$

In this reaction sodium loses an electron (sodium is oxidised) and chlorine gains an electron (chlorine is reduced). Since an oxidation and a reduction take place simultaneously, such reactions are called **redox reactions** (**red** for **red**uction and **ox** for **ox**idation).

An oxidising agent (oxidant) is an electron acceptor and a reducing agent (reductant) is an electron donor. In the above reaction sodium donates an electron and is the reductant. Chlorine accepts an electron and is the oxidant.

The reaction between magnesium and oxygen to form magnesium oxide is another example of a redox reaction. A magnesium atom ($1s^2 2s^2 2p^6 3s^2$) is oxidised to a magnesium ion by losing two electrons and an oxygen atom ($1s^2 2s^2 2p^4$) is reduced to an oxide ion by gaining two electrons. Note that *the oxidant is reduced and the reductant is oxidised during a redox reaction.* The equations for the changes that take place during the reaction are given below.

$$\underset{\text{Reductant}}{Mg} \longrightarrow Mg^{2+} + 2e^-$$

$$\underset{\text{Oxidant}}{\frac{1}{2}O_2} + 2e^- \longrightarrow O^{2-}$$

$$Mg^{2+} + O^{2-} \longrightarrow MgO$$

When metals react with non-metals to form ionic compounds, it is always redox reactions that take place. Metal atoms lose one or more electrons and are oxidised to form positive ions. At the same time non-metal atoms take one or more electrons and are reduced to form negative ions. The number of electrons donated by a metal atom equals the number of valence electrons, and that gained by a non-metal atom equals the number required to complete its outermost orbit.

7.2 Writing partial ionic equations for oxidation and reduction

Let us consider the reaction between aluminium and chlorine to form aluminium chloride. This is a redox reaction.

$$2Al + 3Cl_2 \longrightarrow 2AlCl_3$$

How do we write partial ionic equations for the oxidation and reduction reactions? Here are the steps to follow.

1. Identify the reductant and its oxidised form and write them separated by an arrow. In this case, Al is the reductant and its oxidised form is Al^{3+}.

$$\underset{\text{Reductant}}{Al} \longrightarrow Al^{3+}$$

2. Balance for the total charge on both sides of the equation by adding electrons on the right hand side of the equation. Remember that a reductant donates electrons, in this case, three electrons. This gives the partial equation for oxidation.

$$\underset{\text{Reductant}}{Al} \longrightarrow Al^{3+} 3e^- \text{— partial equation for oxidation}$$

3. Identify the oxidant and its reduced form and write them separated by an arrow.

$$\underset{\text{Oxidant}}{Cl_2} \longrightarrow Cl^-$$

4. Balance for the number of atoms first (in this case, two Cl^- ions from one Cl_2 molecule), and then the number of charges by adding electrons on the left hand side of the equation. (An oxidant adds or accepts electrons.) This gives the partial equation for reduction.

$$\underset{\text{Oxidant}}{Cl_2} + 2e^- \longrightarrow 2Cl^- \text{— partial equation for reduction}$$

Practice question 7.1 Write partial equations for the oxidation of the following atoms or ions:

(a) Na (b) Fe to Fe^{2+} (c) Fe to Fe^{3+} (d) Fe^{2+} to Fe^{3+} (e) Br^-

Practice question 7.2 Write partial equations for the reduction of the following atoms or ions.

(a) Br_2 (b) Fe^{3+} to Fe^{2+} (c) Cu^{2+} to Cu (d) H^+ (e) Al^{3+}

7.3 Overall equation for a redox reaction

A partial equation for oxidation and one for reduction are added to get an overall equation for a redox reaction.

1. First, note the number of electrons in the two partial equations. If they are not the same, equalise the number of electrons by multiplying the equations by appropriate simple whole numbers. In this case, multiply the partial equation for oxidation by 2 and that for reduction by 3.

$$2Al \longrightarrow 2Al^{3+} + 6e^-$$

$$3Cl_2 + 6e^- \longrightarrow 6Cl^-$$

2. Add the two equations to get the overall equation. Cancel out the electrons from the two sides of the equation. There are no electrons in the overall equation indicating that the number of electrons donated and accepted are balanced.

$$\begin{array}{l} 2Al \longrightarrow 2Al^{3+} + \cancel{6}e^- \\ 3Cl_2 + \cancel{6}e^- \longrightarrow 6Cl^- \\ \hline 2Al + 3Cl_2 \longrightarrow 2Al^{3+} + 6Cl^- (\text{or, } 2AlCl_3) \end{array}$$

Another redox reaction

Let us study the reaction between iron (III) chloride and potassium iodide. When aqueous solutions containing these two reactants are mixed, the resultant solution becomes brown due to the formation of iodine.

$$2FeCl_3(aq) + 2KI(aq) \longrightarrow 2FeCl_2(aq) + 2KCl(aq) + I_2(aq)$$

In this reaction, an I^- ion from KI loses an electron and is oxidised to I. Two I atoms pair to form an I_2 molecule. An Fe^{3+} ion from $FeCl_3$ takes an electron and is reduced to an Fe^{2+} ion. I^- ion is the reductant and Fe^{3+} ion is the oxidant. The partial equations for the oxidation of I^- and reduction of Fe^{3+} and the overall equation are given below. Note that the overall equation is obtained by multiplying the equation for reduction by 2 to equalise the number of electrons, and then adding to the equation for oxidation and cancelling the electrons in the final equation.

$$2I^- \longrightarrow I_2 + 2e^- \quad \text{— oxidation}$$

$$Fe^{3+} + e^- \longrightarrow Fe^{2+} \quad \text{— reduction}$$

$$2I^- + 2Fe^{3+} \longrightarrow 2Fe^{2+} + I_2 \quad \text{— redox}$$

Note that the Cl^- ions from $FeCl_3$ and the K^+ ions from KI do not take part in the reaction and they do not appear in the ionic equations.

Example 7.1 Zinc reacts with copper (II) sulphate to give zinc sulphate and copper. Name the oxidant and reductant and write partial equations for the oxidation and reduction and the overall equation for the redox reaction.

Hint: In this reaction zinc is changed to zinc ion (Zn^{2+}) and copper (II) ion (Cu^{2+}) to copper.

Solution

The oxidant is copper (II) ion, (electron acceptor) and the reductant is zinc metal (electron donor).

$$Zn \longrightarrow Zn^{2+} + 2e^- \quad \text{— oxidation}$$
$$Cu^{2+} + 2e^- \longrightarrow Cu \quad \text{— reduction}$$
$$Zn + Cu^{2+} \longrightarrow Zn^{2+} + Cu \quad \text{— redox}$$

Practice question 7.3 Write partial ionic equations for the oxidation and reduction reactions and a balanced overall equation for the reaction between chlorine and potassium bromide solution to give potassium chloride and bromine. Name the oxidant and reductant.

Practice question 7.4 Calcium reacts with bromine to give calcium bromide, $CaBr_2$. Write partial ionic equations for the oxidation and reduction and the overall equation for the redox reaction.

More examples

The examples given above are simple redox reactions. A few, more difficult ones are examined here.

(a) A solution of potassium permanganate ($KMnO_4$) acidified with dilute sulphuric acid reacts with iron (II) sulphate to produce manganese (II) sulphate, iron (III) sulphate and water. In this reaction, the oxidant, MnO_4^- is reduced to Mn^{2+} and the reductant, Fe^{2+} is oxidised to Fe^{3+}. The acid supplies H^+ for the reaction. Here are the steps to follow in order to write the equations.

1. Write the reductant (Fe^{2+}) and its oxidised form (Fe^{3+}) separated by an arrow and balance the equation with electrons.

$$Fe^{2+} \longrightarrow Fe^{3+} + e^- \text{— partial equation for oxidation}$$

2. Write the oxidant (MnO_4^-) and its reduced form (Mn^{2+}) separated by an arrow.

$$MnO_4^- \longrightarrow Mn^{2+}$$

3. Add four molecules of H_2O on the right hand side of the equation to balance for oxygen atoms and then, eight H^+ ions on the left hand side to equalise the hydrogen atoms.

$$MnO_4^- + 8H^+ \longrightarrow Mn^{2+} + 4H_2O$$

4. Balance for the number of charges by adding five electrons on the left hand side. (Mn^{7+} in MnO_4^- to Mn^{2+} requires $5e^-$. Refer to Example 7.2)

$$MnO_4^- + 8H^+ + 5e^- \longrightarrow Mn^{2+} + 4H_2O \text{ — partial equation for reduction}$$

5. To get the overall equation for the redox reaction, multiply the partial equation for oxidation by 5, add to the partial equation for reduction and cancel out the electrons.

$$\underset{\text{Reductant}}{5Fe^{2+}} + \underset{\text{Oxidant}}{MnO_4^-} + 8H^+ \longrightarrow 5Fe^{3+} + Mn^{2+} + 4H_2O \text{ — overall equation}$$

(b) Hydrogen peroxide, H_2O_2, in the presence of an acid, acts as an oxidant and reacts with reductants like potassium iodide. H_2O_2 is reduced to H_2O and I^- is oxidised to I_2.

1. The partial equation for the oxidation of I^- to I_2 is

$$\underset{\text{Reductant}}{2I^-} \longrightarrow I_2 + 2e^- \text{ — oxidation}$$

2. To write the equation for the reduction of H_2O_2 to H_2O, first write,

$$H_2O_2 \longrightarrow H_2O$$

3. Equalise for oxygen atoms by introducing a second H_2O molecule.

$$H_2O_2 \longrightarrow 2H_2O$$

4. Balance for hydrogen atoms by adding H^+ ions on the left hand side of the equation. (The acid medium provides H^+.)

$$H_2O_2 + 2H^+ \longrightarrow 2H_2O$$

5. Add $2e^-$ on the left hand side of the equation to balance the charge. (Remember that reduction is addition of electrons.)

$$\underset{\text{Oxidant}}{H_2O_2} + 2H^+ + 2e^- \longrightarrow 2H_2O \text{ — reduction}$$

6. Finally add the two partial equations to get the overall equation.

$$2I^- + H_2O_2 + 2H^+ \longrightarrow I_2 + 2H_2O \text{ — redox}$$

Redox reactions are not all electron transfer reactions. For example the reaction between sulphur and oxygen to form sulphur dioxide, is similar to that between magnesium and oxygen to form magnesium oxide and both are redox reactions. Electrons are not transferred completely between sulphur and oxygen, but shared in sulphur dioxide. To include such reactions, the concept of oxidation and reduction is extended in terms of change in oxidation number.

7.4 Redox reactions: in terms of change in oxidation number

We need to know how to assign oxidation numbers to atoms (in substances) in order to understand redox reactions in terms of change in oxidation number.

Oxidation number

Oxidation number is the effective number of charges on an atom in a molecule. Oxidation numbers of atoms in substances are calculated using a set of rules.

Rules to assign oxidation numbers

1. All atoms in the elemental state are assigned zero oxidation number. (It is common practice to write the oxidation number of an atom above it in brackets.) For example,

$$\overset{(0)}{H_2},\ \overset{(0)}{Na},\ \overset{(0)}{C},\ \overset{(0)}{Al}$$

2. The oxidation number of hydrogen in most of its compounds is +1, but in metal hydrides it is –1. For example,

$$\overset{(+1)}{H_2}O,\ C\overset{(+1)}{H_4},\ \overset{(+1)}{H_2}SO_4,\ Na\overset{(+1)}{H}CO_3$$

3. The oxidation number of oxygen in most of its compounds is –2, but in peroxides it is –1 and in fluorine oxide, F_2O, it is +2. For example,

$$H_2\overset{(-2)}{O},\ HN\overset{(-2)}{O_3},\ C\overset{(-2)}{O}Cl_2,\ CaC\overset{(-2)}{O_3}$$

4. The oxidation number of a simple ion equals the charge of the ion.

$$\overset{(+1)}{Na^+}\quad \overset{(+3)}{Fe^{3+}},\ \overset{(-1)}{Cl^-},\ \overset{(+2)}{Ca^{2+}}$$

5. In a covalent molecule, the more electronegative element assumes a negative oxidation number. For example, the oxidation number of carbon in methane (CH_4) is –4 since carbon is more electronegative than hydrogen. Intetrachloromethane (CCl_4), the oxidation number of carbon is +4, chlorine being more electronegative than carbon.

$$\overset{(-4)}{C}\overset{(+1)}{H_4},\ \overset{(+4)}{C}\overset{(-1)}{Cl_4},\ \overset{(+2)}{S}\overset{(-1)}{Cl_2},\ \overset{(+2)}{O}\overset{(-1)}{F}$$

6. The sum of the oxidation numbers of all the atoms in a molecule equals zero, and that of an ion equals the total charge of the ion. Note the examples.

$$\overset{(+1)}{H_2}\overset{(-2)}{O},\ \overset{(-4)}{C}\overset{(+1)}{H_4},\ \overset{(+2)}{Mg}\overset{(-1)}{Cl_2},\ \overset{(+1)}{H}\overset{(+5)}{N}\overset{(-2)}{O_3}$$

$$\overset{(+5)}{N}\overset{(-2)}{O_3^-},\ \overset{(-3)}{N}\overset{(+1)}{H_4^+},\ \overset{(+6)}{S}\overset{(-2)}{O_4^{2-}},\ \overset{(+6)}{Cr_2}\overset{(-2)}{O_7^{2-}}$$

Notes

1. In H_2O, the sum of the oxidation numbers of two atoms of hydrogen and one atom of oxygen is zero. In HNO_3, the oxidation number of nitrogen should be +5, so that the sum of the oxidation numbers of one hydrogen, one nitrogen and three oxygen atoms can equal zero.
2. In NO_3^-, the sum of the oxidation numbers of one nitrogen atom (+5) and three oxygen atoms (–6) equals –1, which is the charge of the ion.
3. These rules help us to deduce the oxidation numbers of unknown atoms in molecules or ions.

Example 7.2

Write the oxidation number of each atom in the following compounds or ions:

(i) HCl (ii) $NaHCO_3$ (iii) CO_3^{2-} (iv) $KMnO_4$ (v) $Cu(NO_3)_2$ (vi) MnO_4^-

Solution

(i) $\overset{(+1)}{H}\overset{(-1)}{Cl}$ (ii) $\overset{(+1)}{Na}\overset{(+1)}{H}\overset{(+4)}{C}\overset{(-2)}{O_3}$ (iii) $\overset{(+4)}{C}\overset{(-2)}{O_3^{2-}}$

(iv) $\overset{(+1)}{K}\overset{(+7)}{Mn}\overset{(-2)}{O_4}$ (v) $\overset{(+2)}{Cu}(\overset{(+5)}{N}\overset{(-2)}{O_3})_2$ (vi) $\overset{(+7)}{Mn}\overset{(-2)}{O_4^-}$

Oxidation and reduction in terms of change in oxidation number

Oxidation is an increase in oxidation number and **reduction is a decrease in oxidation number.**

During a redox reaction the oxidation number of an atom is increased (oxidation) and there is a simultaneous decrease in the oxidation number of another atom (reduction). The reductant has its oxidation number increased (i.e. oxidised), and the oxidant its oxidation number decreased (i.e. reduced). This is exemplified in the following reactions.

1. Zinc reacts with dilute hydrochloric acid to produce zinc chloride and hydrogen. The oxidation number of Zn is increased from 0 (in Zn) to +2 (in $ZnCl_2$), that is, Zn is oxidised to $ZnCl_2$. The oxidation number of hydrogen is decreased from +1 (in HCl) to 0 (in H_2), i.e., HCl reduced to H_2. Zn is the reductant and HCl is the oxidant.

$$\underset{\text{Reductant}}{\overset{(0)}{Zn}} + \underset{\text{Oxidant}}{2\overset{(+1)}{H}\overset{(-1)}{Cl}} \longrightarrow \overset{(+2)}{Zn}\overset{(-1)}{Cl_2} + \overset{(0)}{H_2}$$

2. When magnesium is heated with concentrated sulphuric acid, magnesium sulphate, sulphur dioxide and water are formed. In this reaction, the oxidation number of Mg is increased from 0 (in Mg) to +2 (in $MgSO_4$) and that of S is reduced from +6 (in H_2SO_4) to +4 (in SO_2). Mg is the reductant and is oxidised to $MgSO_4$, and H_2SO_4 is the oxidant and is reduced to SO_2.

$$\underset{\text{Reductant}}{\overset{(0)}{Mg}} + \underset{\text{Oxidant}}{2\overset{(+1)}{H_2}\overset{(+6)}{S}\overset{(-2)}{O_4}} \longrightarrow \overset{(+2)}{Mg}\overset{(+6)}{S}\overset{(-2)}{O_4} + \overset{(+4)}{S}\overset{(-2)}{O_2} + 2\overset{(+1)}{H_2}\overset{(-2)}{O}$$

3. The reaction between hydrochloric acid and sodium hydroxide is not a redox reaction since there is no change in the oxidation number of any of the atoms. Similarly, the reaction between sodium carbonate and nitric acid is not a redox reaction.

$$\overset{(+1)}{H}\overset{(-1)}{Cl} + \overset{(+1)}{Na}\overset{(-2)}{O}\overset{(+1)}{H} \longrightarrow \overset{(+1)}{Na}\overset{(-1)}{Cl} + \overset{(+1)}{H_2}\overset{(-2)}{O}$$

$$\overset{(+1)}{Na_2}\overset{(+4)}{C}\overset{(-2)}{O_3} + 2\overset{(+1)}{H}\overset{(+5)}{N}\overset{(-2)}{O_3} \longrightarrow 2\overset{(+1)}{Na}\overset{(+5)}{N}\overset{(-2)}{O_3} + \overset{(+4)}{C}\overset{(-2)}{O_2} + \overset{(+1)}{H_2}\overset{(-2)}{O}$$

It is thus possible to work out whether a reaction is redox or not by noting the oxidation numbers of all atoms in the reactants and products. If there are changes in the oxidation numbers, it is a redox reaction. If there is no change, it is not a redox reaction.

Practice question 7.5 Assign oxidation numbers and decide which of the following are redox reactions.

(a) $NaCl(aq) + AgNO_3(aq) \longrightarrow NaNO_3(aq) + AgCl(s)$

(b) $4Al(s) + 3O_2(g) \longrightarrow 2Al_2O_3(s)$

(c) $Ca(s) + 2H_2O(l) \longrightarrow Ca(OH)_2(aq) + H_2(g)$

(d) $KHCO_3(s) + HCl(aq) \longrightarrow KCl(aq) + CO_2(g) + H_2O(1)$

7.5 Oxidising agents

Below is listed some common oxidising agents.

Name	Formula
Oxygen	O_2
Halogens	F_2, Cl_2, Br_2, I_2
Acidified permanganate, [manganese (VII)]	MnO_4^-, H^+
Acidified dichromate, [chromum(VI)]	$Cr_2O_7^{2-}$, H^+
Iron(III)	Fe^{3+}
Copper(II)	Cu^{2+}
Hydrogen peroxide	H_2O_2
Nitric acid	HNO_3
Concentrated sulphuric acid	H_2SO_4

They vary in oxidising strength. For example, oxygen is a stronger oxidant than bromine, and permanganate is stronger than dichromate. These oxidants accept electrons from reductants during a redox reaction.

A solution of acidified potassium permanganate or dichromate is used in the laboratory to test for reducing agents. Both are powerful oxidants and react with a number of reductants. A solution of permanganate is purple in colour and when it reacts with a reductant, it is reduced to manganese (II) ion which is pale pink (almost colourless in solution). A solution of dichromate is orange in colour and during a reaction of it with a reductant, the colour changes to green due to the formation of chromium (III) ion.

If when a substance is mixed with acidified permanganate, the purple colour disappears, it means that the substance under consideration is a reducing agent. For example, when acidified potassium permanganate solution is mixed with iron (II) sulphate solution, there is decolourisation. MnO_4^- is reduced to Mn^{2+} and Fe^{2+} is oxidised to Fe^{3+}.

$$5Fe^{2+}(aq) + MnO_4^-(aq) + 8H^+(aq) \longrightarrow 5Fe^{3+}(aq) + Mn^{2+}(aq) + 4H_2O(l)$$

7.6 Reducing agents

Some of the common reducing agents are

Name	Formula
Metals	Eg., Na, Mg, Ca, Zn
Hydrogen	H_2
Carbon	C
Carbon monoxide	CO
Hydrogen sulpide	H_2S
Iron (II), ($FeCl_2$)	Fe^{2+}
Iodide, (KI)	I^-

A suitable reducing agent that can be used to test for an oxidising agent in the laboratory is potassium iodide solution. When potassium iodide reacts with an oxidant, potassium iodide is oxidised to iodine while the oxidant is reduced. The solution turns brown due to the formation of iodine.

When a substance is added to potassium iodide solution, a change in colour of the solution to brown indicates that the substance is an oxidant. For example when potassium iodide solution is treated with iron (III) chloride solution, the colour of the solution changes to brown. KI reduces $FeCl_3$ to $FeCl_2$, and itself oxidised to brown I_2.

$$2I^-(aq) + 2Fe^{3+}(aq) \longrightarrow I_2(aq) + 2Fe^{2+}(aq)$$

Tutorial: helping you learn

Progress questions

7.1 Write partial equations for the oxidation of the following atoms or ions.
(a) Ca (b) K (c) Cu^+ to Cu^{2+} (d) Pb to Pb^{2+} (e) H^-

7.2 Write partial equations for the reduction of the following atoms or ions.
(a) S (b) I_2 (c) Fe^{3+} to Fe (d) Cu^{2+} to Cu^+ (e) Ag^+

7.3 Assign oxidation numbers to all the elements in the following substances.
(a) $FeCl_2$ (b) $NaNO_3$ (c) $CuSO_4$ (d) $KHCO_3$ (e) NH_4Cl

7.4 Assign oxidation numbers to all the elements in the following ions.
(a) MnO_4^- (b) SO_4^{2-} (c) NH_4^+ (d) CO_3^{2-} (e) NO_3^-

7.5 (a) Assign oxidation numbers and decide which of the following are redox-reactions.

$$\text{(i) } Ca(s) + 2HCl(aq) \longrightarrow CaCl_2(aq) + H_2(g)$$
$$\text{(ii) } Cl_2(g) + 2KI(aq) \longrightarrow I_2(aq) + 2KCl(aq)$$
$$\text{(iii) } Ca(OH)_2(s) + 2HCl(aq) \longrightarrow CaCl_2(aq) + 2H_2O(l)$$
$$\text{(iv) } Fe(s) + 2AgNO_3(aq) \longrightarrow Fe(NO_3)_2(aq) + 2Ag(s)$$

(b) Name the oxidant and reductant in each of the above redox reactions.

7.6 Assigning oxidation numbers for all the atoms, find the oxidant and reductant for each of the following reactions. Write partial equations for oxidation and reduction and the overall equation for the redox reaction.

$$\text{(i) } 2Fe(s) + 3Cl_2(g) \longrightarrow 2FeCl_3(s)$$
$$\text{(ii) } Cl_2(g) + 2KI(aq) \longrightarrow 2KCl(aq) + I_2(aq)$$
$$\text{(iii) } Cl_2(g) + 2NaOH(aq) \longrightarrow NaCl(aq) + NaOCl(aq) + H_2O(l)$$
$$\text{(iv) } Cu(s) + 2AgNO_3(aq) \longrightarrow 2Ag(s) + Cu(NO_3)_2(aq)$$
$$\text{(v) } Mg(s) + FeCl_2(aq) \longrightarrow Fe(s) + MgCl_2(aq)$$
$$\text{(vi) } Ca(s) + H_2SO_4(aq) \longrightarrow CaSO_4(aq) + H_2(g)$$
$$\text{(vii) } Cr_2O_7^{2-}(aq) + 14H^+(aq) + 6Fe^{2+}(aq) \longrightarrow 2Cr^{3+}(aq) + 6Fe^{3+}(aq) + 7H_2O(l)$$
$$\text{(viii) } 2MnO_4^-(aq) + 5H_2O_2(aq) + 6H^+(aq) \longrightarrow 2Mn^{2+}(aq) + 5O_2(g) + 8H_2O(l)$$

8 Introduction to Organic Chemistry

One-minute summary

Organic chemistry deals with the study of carbon compounds. In this chapter we shall briefly appraise a few classes of organic compounds, namely; alkanes, alkenes, halogenoalkanes and alcohols. Each of the above classes forms an homologous series of compounds. The different members of a homologous series possess similar properties and so our discussion is limited to the first few members of each series. We will come across a number of different reaction types in this chapter. These include oxidation, substitution, addition, elimination and polymerisation. We shall also study the structure and bonding, molecular and structural formulae and IUPAC nomenclature of organic compounds.

The highlights of the chapter are:

- structure and bonding
- molecular and structural formulae
- isomerism
- IUPAC nomenclature
- properties and reactions of alkanes
- alkenes
- halogenoalkanes
- alcohols

8.1 Representing formulae of organic compounds

There are tens of thousands of organic compounds known to scientists and the study of these compounds is simplified by the study of the classes to which these compounds belong. The simplest class of organic compounds is the hydrocarbons i.e. compounds containing carbon and hydrogen. Hydrocarbons can be classified as **saturated** and **unsaturated**, and also as **aliphatic** and **aromatic** hydrocarbons.

In **saturated hydrocarbons**, each carbon atom is bonded to four other atoms by single covalent bonds and in **unsaturated hydrocarbons** at least two carbon atoms in a molecule are bonded to each other by a double bond (two covalent bonds) or a triple bond (three covalent bonds). Note these examples.

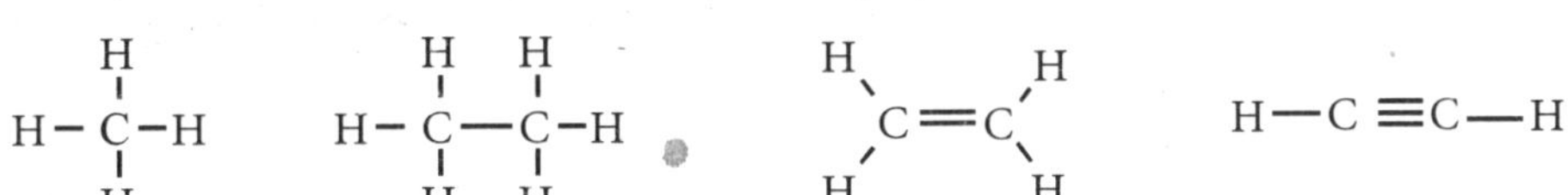

Methane, Ethane — Saturated hydrocarbons

Ethene, Ethyne (Acetylene) — Unsaturated hydrocarbons

Aliphatic compounds may be cyclic or open-chained compounds. All the compounds pictured above are examples of open-chained aliphatic hydrocarbons. **Aromatic** compounds are those which contain one or more benzene ring.

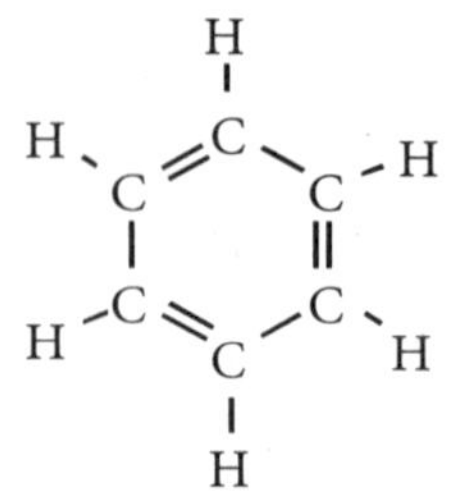

Benzene

An aromatic hydrocarbon

Aliphatic hydrocarbons can be further divided into **alkanes**, **alkenes**, **alkynes** and **cycloalkanes**. We shall study these classes of compounds in the latter sections.

Empirical, molecular and structural formulae

The **molecular formula** of a compound displays the type and number of all the atoms in a molecule of it. The **empirical formula** outlines the ratio of the number of atoms in a molecule of the compound (Section 2.8). We have seen in Section 2.8 that the molecular formula can be the empirical formula itself or a simple multiple of the empirical formula.

Molecular formula = $(\text{Empirical formula})_n$

Practice question 8.1 Write the empirical formula of each of the following compounds.

(a) Glucose, $C_6H_{12}O_6$ (b) Propene, C_3H_6 (c) Copper (I) iodide, Cu_2I_2
(d) Methylamine, CH_5N (e) Butane, C_4H10 (f) Pentene, C_5H_{10}

The **structural** formula of a compound is a pictorial representation of how the atoms in a molecule are bonded. Methane is the simplest organic compound. We have seen in Chapter 3 that methane molecule has a tetrahedral shape and the angle between any two C—H bonds in methane is approximately 109.5°. The structural formula of methane is given below.

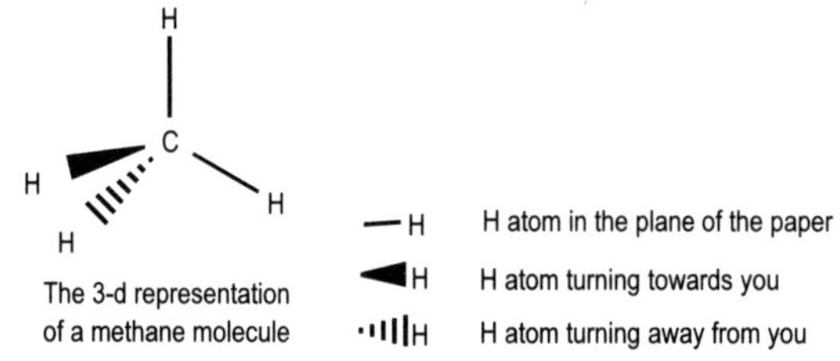

Figure 8.1: Structural formula of methane

It is difficult and time consuming to depict the three-dimensional structures of larger molecules. A simple way of drawing the structure of methane is shown below. It should be remembered that the molecule is not flat, but three-dimensional and the H—C—H bond angles are 109.5° and not 90°.

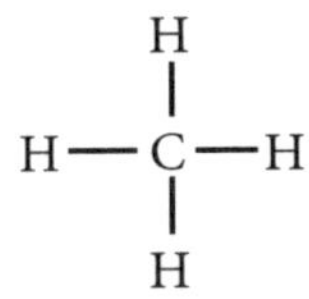

Methane

Note the structural formulae of the following compounds.

Table 8.1: Molecular, empirical and structural formulae of some compounds

Compound	Molecular formula	Empirical formula	Structural formula
Water	H_2O	H_2O	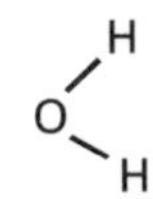
Methane	CH_4	CH_4	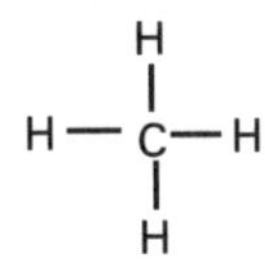
Ethane	C_2H_6	CH_3	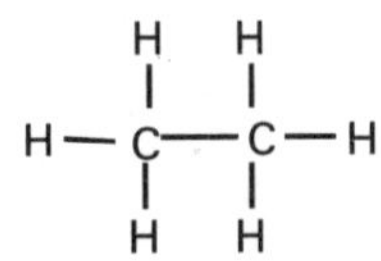
Ethanoic acid	$C_2H_4O_2$	CH_2O	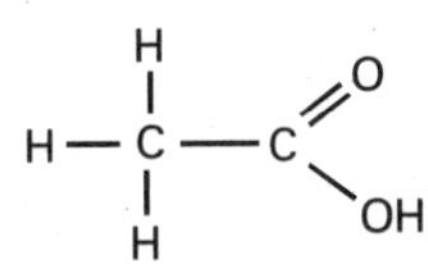

8.2 Alkanes: bonding, structure and isomerism

Alkanes can be considered as the fundamental class of organic compounds and many other classes of compounds are derived from alkanes. Alkanes form a homologous series of compounds.

What is a homologous series? A homologous series is a series of compounds with similar chemical properties and gradually varying physical properties; all the members of which can be represented by a general formula, and each member in the series differs from the next by a $-CH_2-$ group.

The general formula of the alkanes is C_nH_{2n+2}, where n is the number of carbon atoms in the molecule. The first member of the alkanes is methane, CH_4. The names of all alkanes end in *-ane*. The names and formulae of the first six alkanes are given in Table 8.2.

Alkyl groups

Alkyl groups are units derived from alkanes by removing a hydrogen atom and they form as part of some molecules. The name of an alkyl group is derived from the corresponding alkane by substituting the *-ane* ending of the alkane by *-yl*. Thus, methyl group, CH_3–, corresponds to methane, CH_4, ethyl group, C_2H_5–, to ethane, C_2H_6 and so on.

Table 8.2: Alkanes

Name	Formula
Mehane	CH_4
Ethane	C_2H_6
Propane	C_3H_8
Butane	C_4H_{10}
Pentane	C_5H_{12}
Hexane	C_6H_{14}

Table 8.3: Some alkyl groups

Alkane	Formla	Alkyl goup	Formula
Methane	CH_4	Methyl	CH_3–
Ethane	C_2H_6	Ethyl	C_2H_5–
Propane	C_3H_8	Propyl	C_3H_7–

Bonding and structure

Methane is a covalent molecule. The carbon atom ($1s^2\ 2s^2\ 2p^2$) which has four electrons in its outer orbit uses these electrons to share with the electrons of four hydrogen atoms to form four covalent bonds.

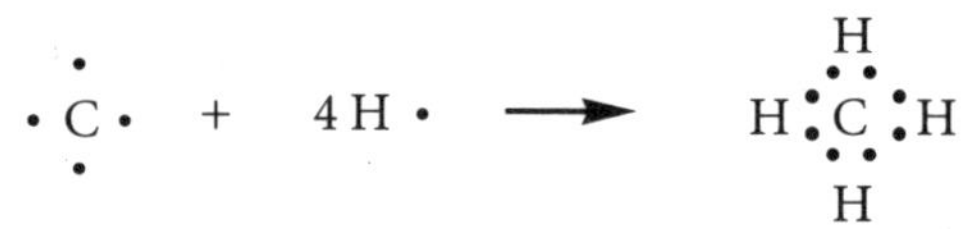

These four shared electron pairs arrange themselves or repel each other so that they are as far away as possible from each other and that happens when they are arranged tetrahedrally. So a molecule of methane is tetrahedral in shape with a carbon atom at the centre of the tetrahedron and hydrogen atoms at the four corners.

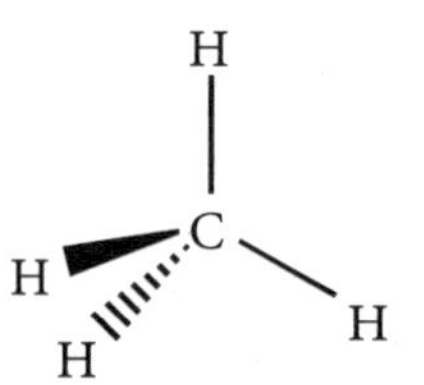

As mentioned in Section 8.1, the structural formula is simplified as shown below for methane and ethane.

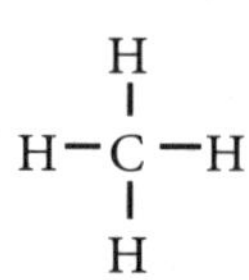

Methane

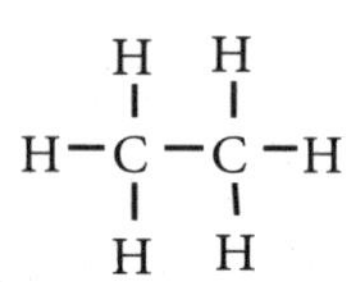

Ethane

A **condensed structural formula** is often used, where each carbon and the hydrogen atoms attached to it are written as a unit. The elaborate and the condensed structural formulae of some compounds are outlined below.

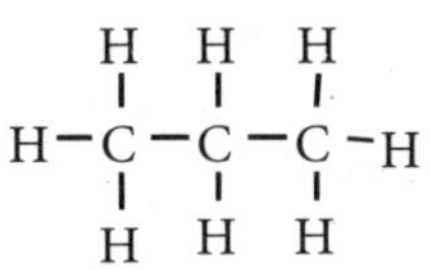

$CH_3CH_2CH_3$

Propane

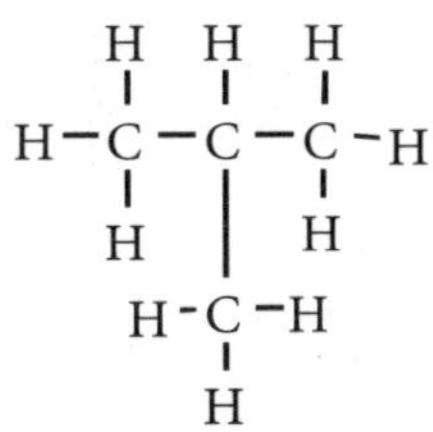

CH_3CHCH_3 (with CH_3 on the middle carbon)

Methylpropane

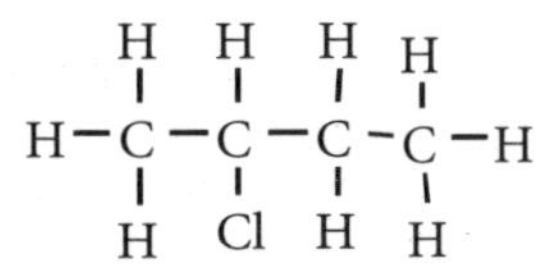

$CH_3CHClCH_2CH_3$

2-Chlorobutane

Isomerism

***Isomers** are compounds with the same molecular formula but different structural formulae, hence different compounds.* There is only one compound each with the molecular formula CH_4, C_2H_6 and C_3H_8. But we can arrange the carbon and hydrogen atoms of butane, C_4H_{10}, in two different ways as shown below.

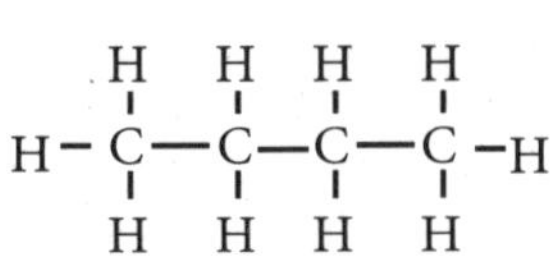

n-Butane

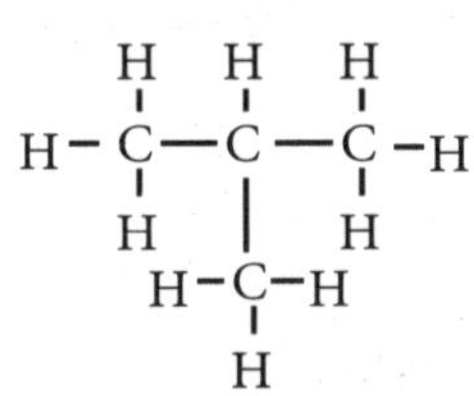

iso-Butane (Methylpropane)

Isomers of butane

These are the isomers of butane; *n*-butane an unbranched alkane and *iso*-butane (methyl propane) a branched alkane. These are two different compounds with the same molecular formula, C_4H_{10}. The number of isomers increases as the number of carbon atoms in the alkane increases. There are three pentanes, five hexanes and so on.

Example 8.1 Write the structural formulae and the condensed structural formulae of all the isomers of pentane, C_5H_{12}.

Solution

(1)

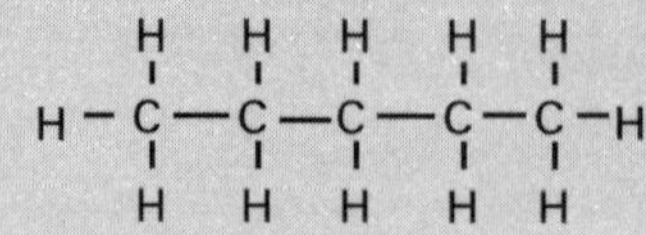

$CH_3CH_2CH_2CH_2CH_3$

(2)

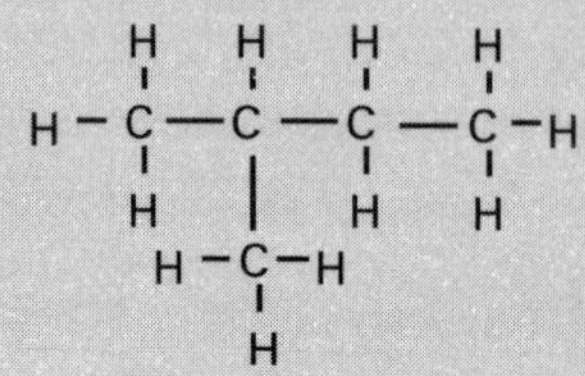

$CH_3CHCH_2CH_3$
  |
  CH_3

(3)

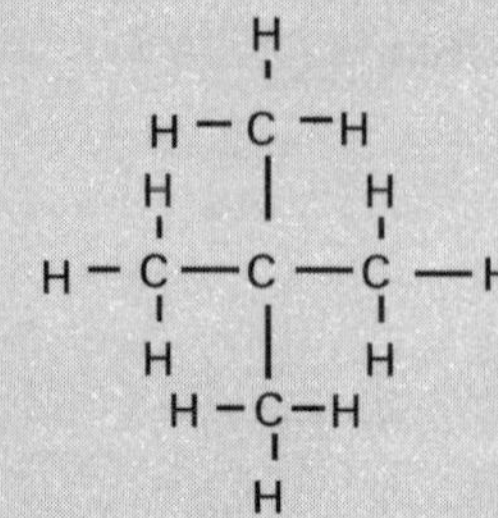

CH_3
|
CH_3 C CH_3
|
CH_3

8.3 Introducing IUPAC nomenclature

There are so many organic compounds that it is impractical to give them individual names. Though some organic compounds are known by their common names (trivial names), it is necessary to name the compounds systematically by a set of rules. The system of naming compounds is called the IUPAC (*International Union of Pure and Applied Chemistry*) nomenclature.

Rules to name alkanes by the IUPAC system

1. The names butane, pentane, hexane etc. are used for the unbranched alkanes. The IUPAC name for $CH_3CH_2CH_2CH_3$ is butane. Its common name is *n*-butane (*normal*-butane).
2. In the case of branched compounds, note the number of carbon atoms in the longest continuous carbon-carbon chain. The parent name is deduced from this chain.

   $CH_3CH_2CHCH_3$ (with CH_3 attached to CH) $CH_3CH_2CHCH_2CH_3$ (with CH_3 attached to CH)

   The names of the above compounds are derived from butane (the four carbon alkane) and pentane (the five carbon alkane) respectively.
3. Number the carbon atoms in the longest chain, starting from the end closer to the branch. Name each substituent and assign a number to locate it in the main chain.

   4 3 2 1: $CH_3CH_2CHCH_3$ (with CH_3 attached to C-2)
   2-methylbutane

   1 2 3 4 5: $CH_3CH_2CHCH_2CH_3$ (with CH_3 attached to C-3)
   3-methylpentane

4. When there are two or more substituents, give each substituent a number to locate it and list them alphabetically. When two or more identical substituents are present, use the prefix di-, tri- etc. with the name. When there are two substituents attached to the same carbon atom, use the number to locate each substituent. Note the following examples.

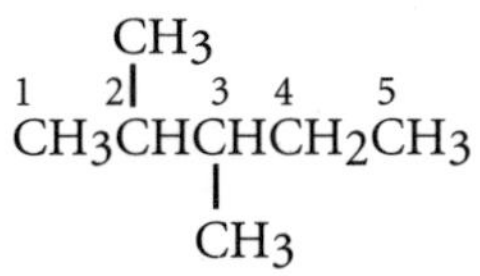
2,3-dimethylpentane

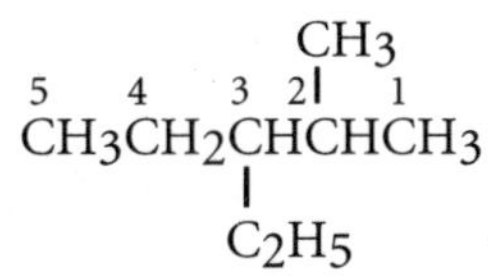
3-ethyl-2-methylpentane

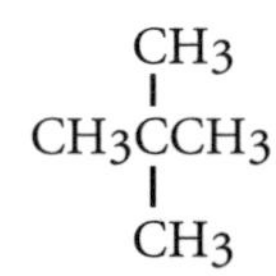
3,3-dimethylpropane

Example 8.2

Write IUPAC names for these alkanes

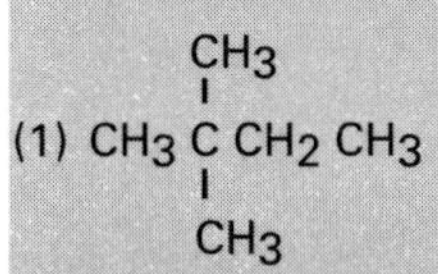

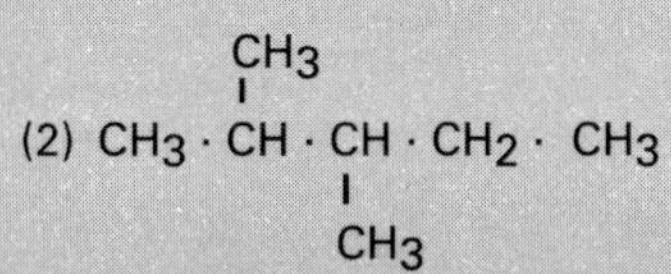

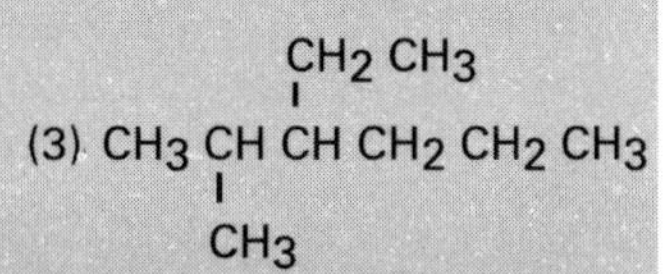

Solution

(1) 2,2-dimethylbutane (2) 2,3-dimethylpentane (3) 3-ethyl-2-methylhexane

8.4 Alkanes: properties and reactions

The first four alkanes are gases and the higher alkanes are either liquids or waxy solids. The intermolecular forces of attraction in alkanes are van der Waals forces and their strength increases with increase in molecular weight. Therefore the melting and boiling points of alkanes increase as we move higher in the series.

Alkanes are insoluble in water. Since the intermolecular forces in water are fairly strong hydrogen bonds and those in alkanes weak van der Waals forces, water molecules cannot form strong bonds with alkane molecules. So alkanes do not dissolve in water. But alkanes are soluble in, or miscible with organic solvents such as trichloromethane, tetrachloromethane and diethyl ether. This is because similar bonds (van der Waals forces) are formed between alkane molecules and these solvent molecules.

Reactions of alkanes

Alkanes are not a very reactive type of compounds. Alkanes undergo two types of reaction: combustion and substitution.

Combustion

Alkanes burn in air or oxygen to form carbon dioxide and water. These are exothermic reactions, producing a lot of heat. Butane is the main constituent of domestic gas and it produces a large quantity of heat during burning.

$$\underset{\text{Methane}}{CH_4(g)} + 2O_2(g) \longrightarrow CO_2(g) + 2H_2O(l)$$

$$\underset{\text{Butane}}{2C_4H_{10}(g)} + 13O_2(g) \longrightarrow 8CO_2(g) + 10H_2O(l)$$

Substitution – Halogenation

When methane and chlorine are mixed in the presence of heat or light, chloromethane and hydrogen chloride are formed.

$$\underset{\text{Methane}}{CH_4(g)} + Cl_2(g) \longrightarrow \underset{\text{Chloromethane}}{CH_3Cl(g)} + HCl(g)$$

In this reaction, a hydrogen atom of methane is replaced by a chlorine atom. Such a reaction is called a *substitution reaction*. The reaction in which chlorine is substituted is called *chlorination*.

If excess chlorine is present, all the hydrogen atoms can be substituted one after the other according to the following equations.

$$CH_3Cl + Cl_2 \longrightarrow \underset{\text{Dichloromethane}}{CH_2Cl_2} + HCl$$

$$CH_2Cl_2 + Cl_2 \longrightarrow \underset{\text{Trichloromethane}}{CHCl_3} + HCl$$

$$CHCl_3 + Cl_2 \longrightarrow \underset{\substack{\text{Tetrachloromethane} \\ \text{(Carbon tetrachloride)}}}{CCl_4} + HCl$$

The reaction mechanism (the reaction pathway)

How does the reaction take place? The reaction mechanism can be explained as follows. A few chlorine atoms are needed to start the reaction. These are formed when some chlorine molecules absorb energy and dissociate into atoms (1). Light or heat provides the energy needed for the dissociation of the bond.

$$Cl{-}Cl \longrightarrow :\ddot{\underset{..}{Cl}}\cdot + :\ddot{\underset{..}{Cl}}\cdot \qquad (1)$$

Each chlorine atom contains seven electrons in its outer orbit and is very reactive since it has a tendency to acquire one more electron in order to complete a stable octet. So a chlorine atom abstracts a hydrogen atom from a methane molecule producing hydrogen chloride. A methyl radical is produced simultaneously (2). (Remember that a covalent bond consists of two electrons. One electron goes with hydrogen and one remains with the carbon of the methyl radical.) This is the *initiation step* (2).

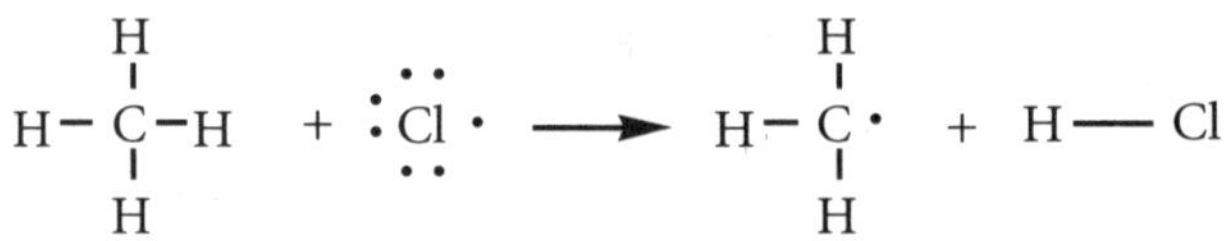

(2) Initiation

Methyl radical

The methyl radical is called a *free radical* and it is very reactive due to the presence of a single electron on the carbon atom. It combines with another chlorine molecule to form chloromethane and a chlorine atom (3). A chain reaction sets in as the chlorine atoms in step (3) react with more methane (2), and so on. This is called the *propagation step*.

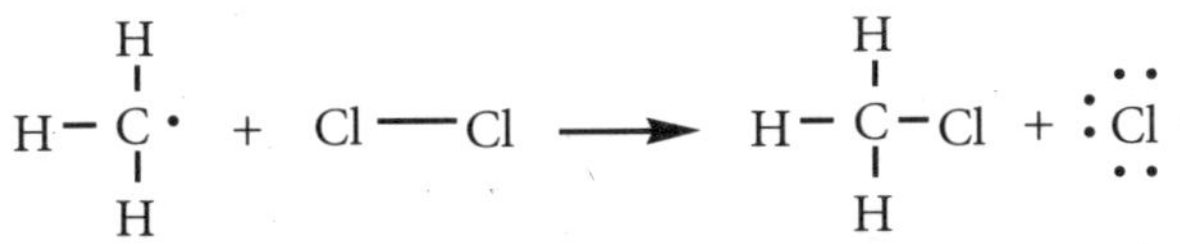

(3) Propagation

The reaction terminates when a methyl radical combines with a chlorine atom (4). This is the *termination step*.

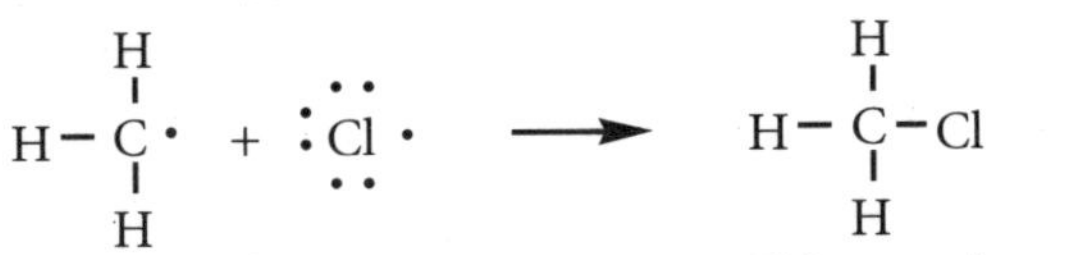

Chloromethane

(4) Termination

This type of substitution reaction is called *free radical substitution* since free radicals take part in the reaction.

Ethane reacts with chlorine in a similar way. Similarly, alkanes react with bromine to form bromine substituted compounds, but the reactions take place at a higher temperature.

8.5 Unsaturated hydrocarbons: alkenes and alkynes

Alkenes and alkynes are *unsaturated* compounds, alkenes containing one double bond per molecule, and alkynes one triple bond. Ethene is the simplest alkene and ethyne, the simplest alkyne.

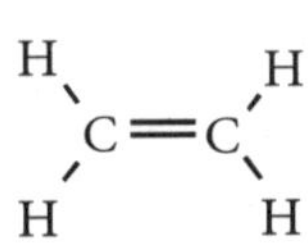

Ethene

$$H-C\equiv C-H$$

Ethyne
(Acetylene)

We shall study the reactions of alkenes in some detail in this chapter. Alkynes resemble alkenes in some of their properties. Both types of compounds undergo addition reactions. When an alkyne molecule adds a molecule of hydrogen, the corresponding alkene is formed. Similarly, alkene adds a molecule of hydrogen to form an alkane.

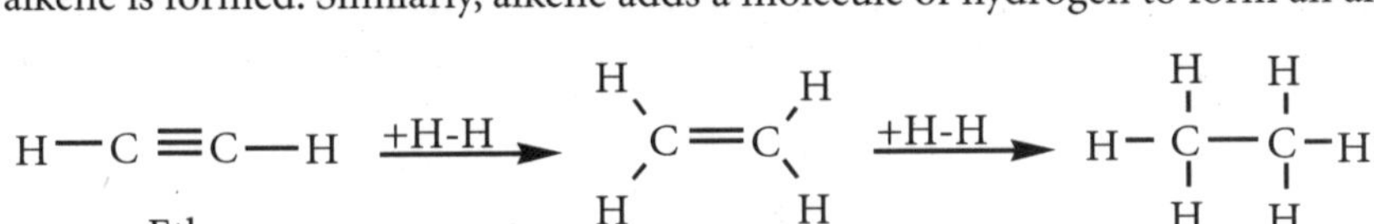

Ethyne (Acetylene) — Ethene — Ethane

$$\text{Alkyne} \xrightarrow{+H_2} \text{Alkene} \xrightarrow{+H_2} \text{Alkane}$$

Alkenes: Nomenclature, bonding and structure

Alkenes form a homologous series of compounds with the general formula C_nH_{2n}, where *n* stands for the number of carbon atoms. The simplest alkene has two carbon atoms in the molecule. The IUPAC names of alkenes are derived from the corresponding alkanes by replacing the *-ane* ending with *-ene*. The names and molecular formulae of the first few members are displayed in Table 8.4

Table 8.4

Alkane	Formula	Alkene	Formula
Methane	CH_4		
Ethane	C_2H_6	Ethene	C_2H_4
Propane	C_3H_8	Propene	C_3H_6
Butane	C_4H_{10}	Butene	C_4H_8
Pentane	C_5H_{12}	Pentene	C_5H_{10}
Hexane	C_6H_{14}	Hexene	C_6H_{12}

Bonding in ethene

There are two carbon atoms in a molecule of ethene. Each carbon atom uses two of its four outer electrons to form covalent bonds with two hydrogen atoms. The remaining two electrons are shared with two electrons of the other carbon atom to form two covalent bonds (a double bond). Therefore, each carbon atom has four bonds, two with two hydrogens and two with the second carbon. The Lewis structure and the structural formula of ethene are shown below.

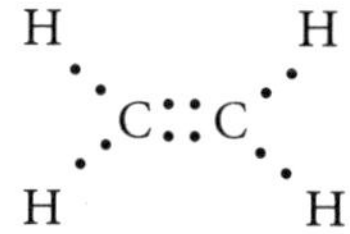

Lewis structure of ethene

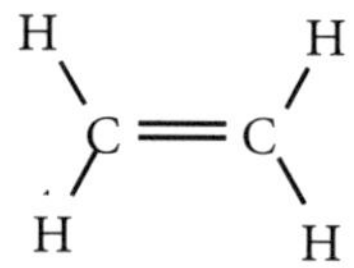

Structural formula of ethene

$CH_2{=}CH_2$

Condensed structural formula of ethene

Isomerism in alkenes

Ethene and propene only have one isomer each. The formula of propene, C_3H_6 is displayed below. Butene, C_4H_8, has three structural isomers. Structures (1) and (2) are straight chain compounds which differ in the position of the double bond and (3) is a branched chain alkene.

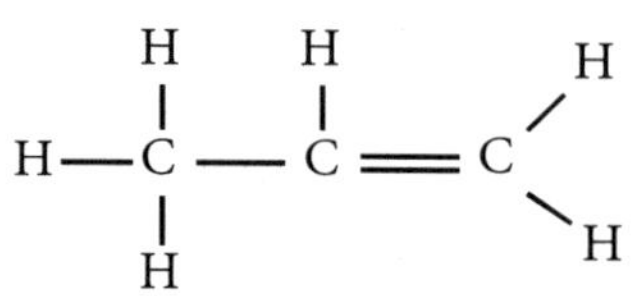

or $CH_3CH{=}CH_2$

Propene

$CH_2{=}CH\,CH_2\,CH_3$

(1)

$CH_3\,CH{=}CH\,CH_3$

(2)

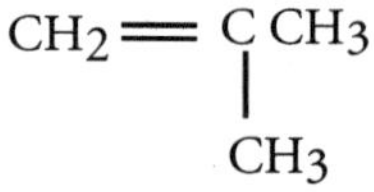

(3)

Isomers of butane

IUPAC rules to name alkenes

1. Find the longest carbon-carbon chain which contains the double bond. The parent name is assigned on the basis of this chain. For example, structure (3) is named from propene and (4) from butene.

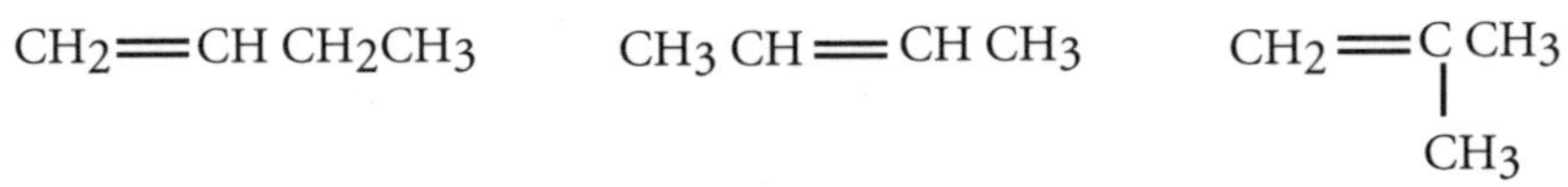

(1) (2) (3)

2. Number the carbon atoms in the chain starting from the end closer to the double bond. If the double bond is placed at an equal distance from both ends, then start numbering the chain closer to a branch. These points are exemplified in structures (4) and (5).

(4) (5)

3. To assign the position of the double bond, use the lower of the two numbers of the carbon atoms which share the double bond as indicated below.

But-1-ene But-2-ene

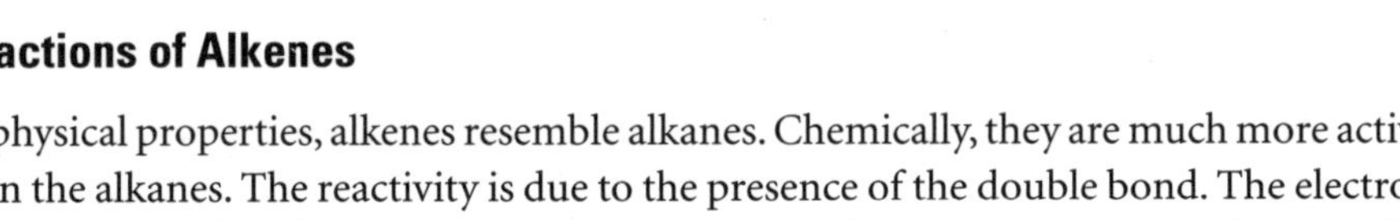

2-Methylpropene 3-Methylbut-1-ene 2-Methylbut-2-ene

Reactions of Alkenes

In physical properties, alkenes resemble alkanes. Chemically, they are much more active than the alkanes. The reactivity is due to the presence of the double bond. The electron density at the double bond is high which makes alkenes to react with electrophilic (electron liking) reagents. The main types of reaction of alkenes are combustion, electrophilic addition and polymerisation.

Note: An **electrophile** (electron liking) is a molecule or a positive ion which can accept a pair of electrons from a donor to form a co-ordinate bond. Molecules such as $AlCl_3$ and ions such as H^+ and Br^+ are examples of electrophiles. Al in $AlCl_3$ molecule has six electrons, H^+ has no electrons and Br^+ has six electrons in the outer orbit. They can accept a pair of electrons by sharing with a donor molecule.

1. Combustion

Like alkanes, alkenes burn in air or oxygen to form carbon dioxide and water.

$$C_2H_4(g) + 3O_2(g) \longrightarrow 2CO_2\,(g) + 2H_2O(l)$$

Practice question 8.2 Write equations for the combustion of propene, C_3H_6 and butene, C_4H_8.

2. Addition reactions

Alkenes react with a number of electrophilic substances such as halogens, hydrogen halides and water to form addition products.

Addition of hydrogen halides

An alkene molecule adds a molecule of hydrogen halide to form a halogenoalkane. For example, ethene reacts with hydrogen bromide to give bromoethane.

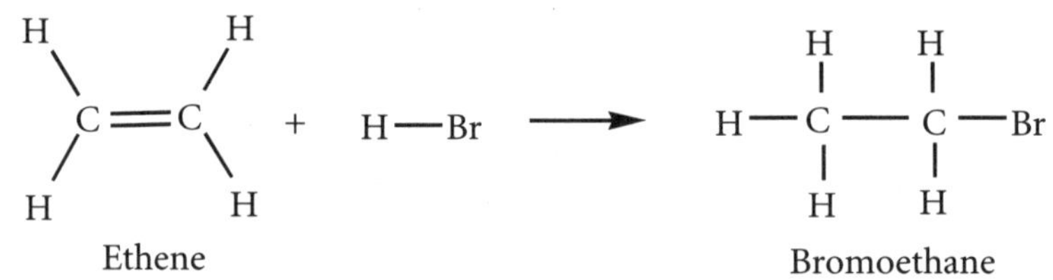

The mechanism of the reaction The electron density at the double bond is high and can attract positively charged particles. Hydrogen bromide is a polar, covalent molecule. Since bromine is more electronegative than hydrogen, the hydrogen end of H—Br is partially positively charged (δ+) and the bromine end partially negatively charged (δ–). When ethene and HBr are mixed, the electrons at the double bond attract the positive hydrogen of H—Br and one of the carbon atoms forms a bond with the hydrogen using a pair of electrons at the double bond. This results in the formation of a carbocation (an ion with a positive charge on the carbon). At the same time, the electron pair between hydrogen and bromine moves to bromine to form a negative bromide ion (Step 1).

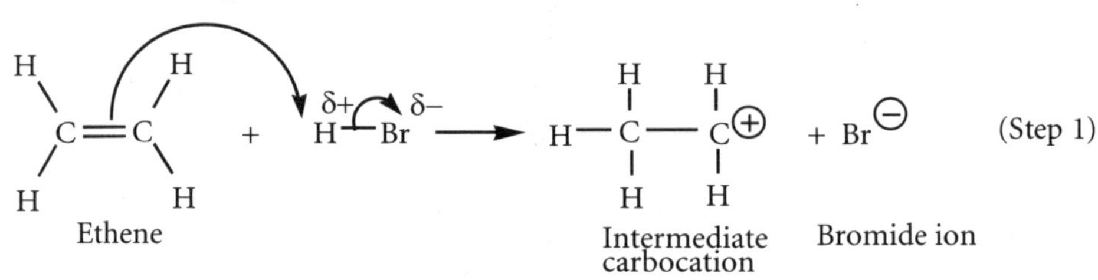

Note: The movement of electrons are shown by curved arrows.

In the second step, the reactive carbocation and the negative bromide ion add to form the product bromoethane.

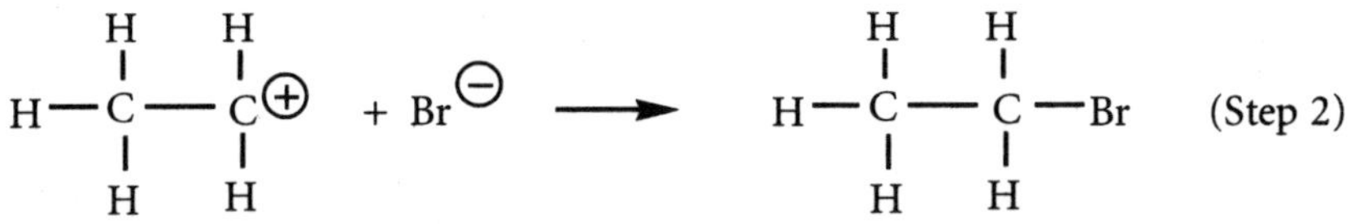

Bromoethane

Ethene reacts with hydrogen chloride in a similar way to from chloroethane.

Propene reacts with hydrogen chloride to form 2-chloropropane and not 1-chloropropane.

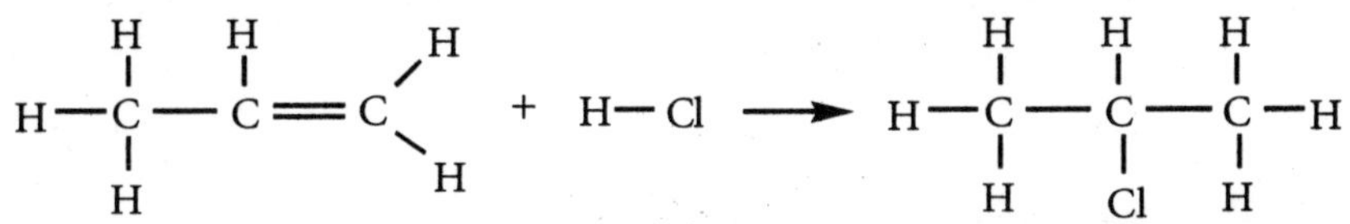

Propene 2-chloropropane

How do we explain the formation of 2-chloropropane in preference to 1-chloropropane? We have seen that the first step in the addition of hydrogen halide is the addition of a hydrogen ion (H^+) to one of the two carbons of the double bond. It can be to C—1 or C—2 of propene. If H^+ is added to C—1, the intermediate carbocation (1) is formed and if it is added to C—2, the carbocation (2).

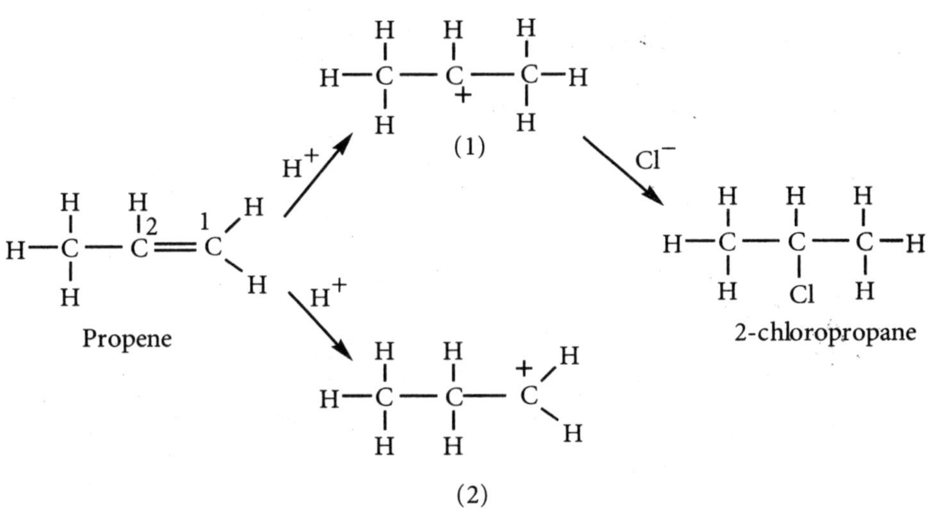

The *secondary carbocation (1) is more stable than the primary carbocation (2)* and so (1) is formed in preference to (2). The addition product of (1) and a chloride ion in the second step is 2-chloropropane.

Note: A primary carbocation is an ion which has a positive charge on a carbon which is bonded to two hydrogen atoms and an alkyl group. A secondary carbocation has one hydrogen and two alkyl groups bonded to the positive carbon and a tertiary carbocation has three alkyl groups bonded to the positive carbon.

The product formed in such a reaction can be predicted using *Markovnikov's rule* which states that when a hydrogen halide is added to an alkene, the hydrogen adds to the carbon which is bonded to more hydrogen atoms.

Example 8.3 What is the main product of the reaction between but-1-ene and HBr? Write an equation for the reaction.

Solution

The main reaction product is 2-bromobutane. H^+ is added on to C—1 to form a secondary carbonium ion, $CH_3CH_2CH^+CH_3$ which adds Br^- to give the product.

$$CH_3CH_2CH{=}CH_2 + HBr \longrightarrow CH_3CH_2\underset{\displaystyle Br}{\underset{|}{C}}HCH_3$$

But-1-ene — 2-Bromobutane

Addition of H_2SO_4

Ethene is absorbed by concentrated sulphuric acid to form ethyl hydrogen sulphate. This is an addition reaction similar to the addition of hydrogen halides.

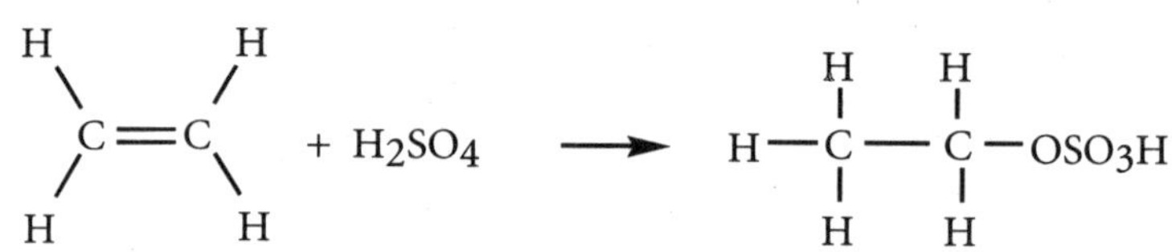

Ethene — Ethyl hydrogen sulphate

Ethyl hydrogen sulphate is hydrolysed in the presence of dilute sulphuric acid to give ethanol.

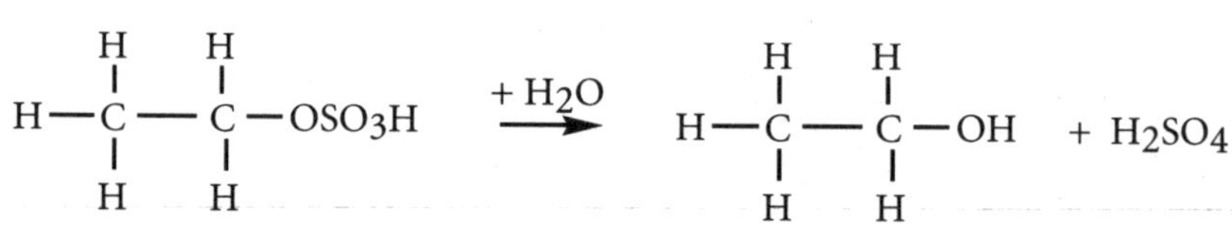

Ethyl hydrogen sulphate — Ethanol

Addition of water (Hydration)

An alkene can add a molecule of water in the presence of an acid catalyst to form an alcohol. This reaction in which water is added to a substance is called **hydration**. When ethene reacts with water in the presence of dilute sulphuric acid as catalyst, ethanol is formed.

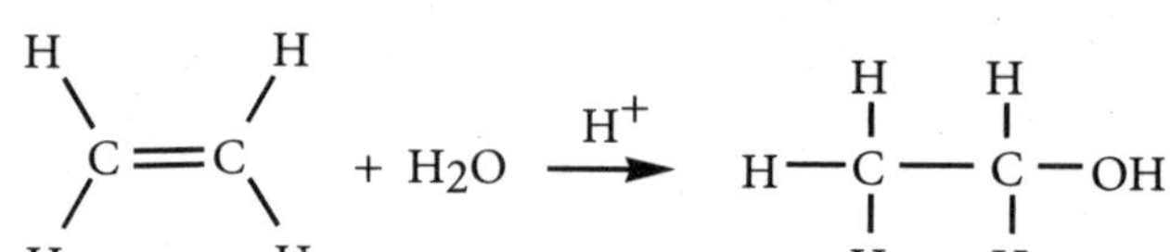

Ethene — Ethanol

Addition of halogens

Chlorine and bromine react with alkenes to form dihalides. Thus chlorine adds on ethene to give 1,2-dichloroethane.

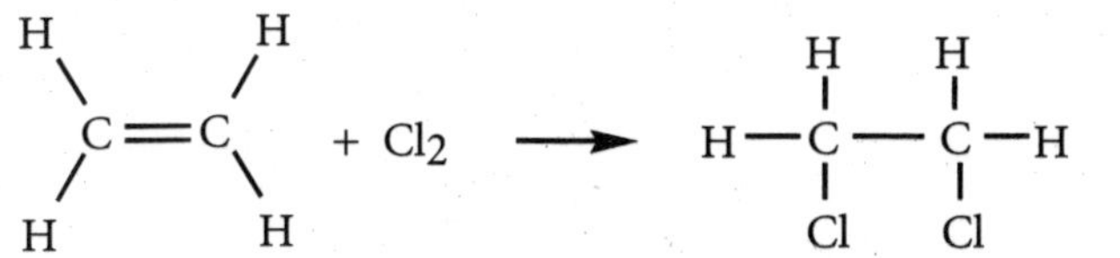

Ethene — 1,2-Dichloroethane

Bromine reacts with alkenes in a similar way, to produce dibromo-derivatives. Bromine water is used conveniently in the laboratory to test for unsaturated compounds. Bromine water is brown and when it is mixed with an unsaturated compound, its brown colour disappears due to the formation of colourless products.

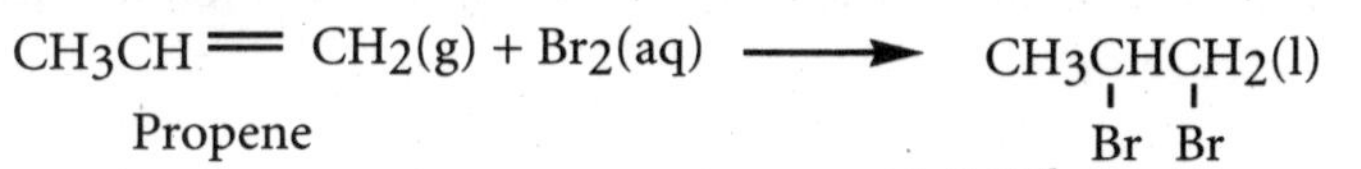

Propene — 1,2-Dibromopropane

Addition of hydrogen

Alkenes add hydrogen in the presence of nickel or platinum catalysts at high temperatures and pressures to form the corresponding alkanes. The addition of hydrogen to an unsaturated compound is called **hydrogenation**. Ethene on hydrogenation is converted to ethane and propene to propane.

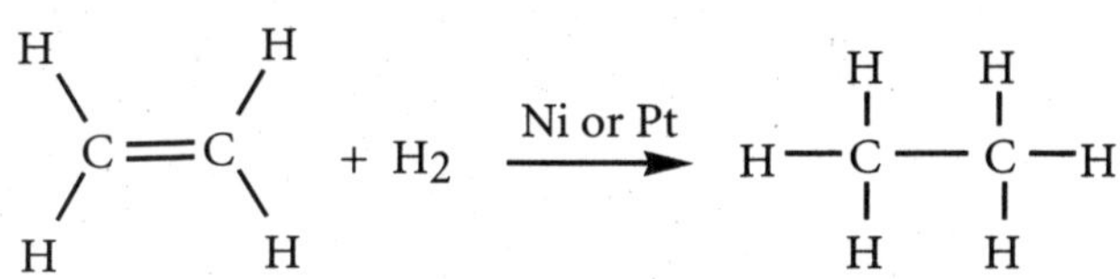

Ethene — Ethane

Hydrogenation is an important reaction for the industrial preparation of margarine. Oils are naturally occurring liquids and chemically they are esters of glycerol (1,2,3-trihydroxypropane) with long chain unsaturated carboxylic acids. When oils are hydrogenated under high pressure and temperature in the presence of a nickel powder catalyst, hydrogen is added at the double bonds to produce corresponding saturated compounds. The product (margarine) after hydrogenation has a higher melting point and is solid.

3. Oxidation

When an alkene is treated with alkaline potassium permanganate, a dihydroxy compound is formed by the addition of two hydroxyl groups on to the double bonded carbon atoms. For example, ethene produces 1,2-dihydroxyethane (ethane-1,2-diol). Potassium permanganate is reduced to brown manganese dioxide at the same time.

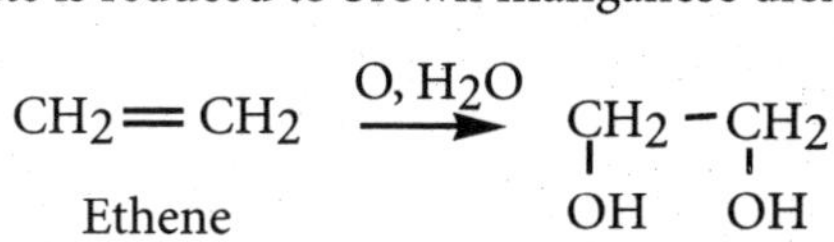

Ethane-1,2-diol

4. Polymerisation

Alkenes undergo polymerisation reactions. For example, ethene polymerises to polyethene (polythene) in the presence of catalysts like benzoyl peroxide. In the polymerisation reaction many molecules of ethene bond to each other to form very large molecules.

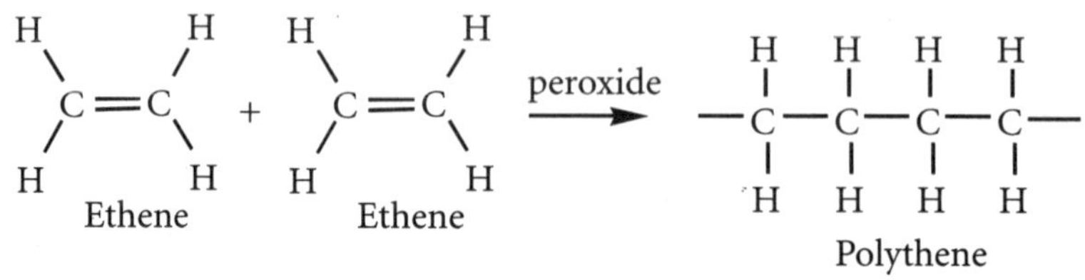

> **Practice question 8.3** Teflon is a polymer used for non-stick coatings and is prepared starting from tetrafluoroethene. Give an equation for the reaction using three monomer units.

8.6 Halogenoalkanes

Halogenoalkanes are compounds formed when one or more hydrogen atoms of an alkane are substituted by halogen atoms. The names of halogenoalkanes can be obtained by adding the prefix fluoro-, chloro-, bromo- and iodo- to the name of the parent compound. The position of the halogen in the molecule is noted by a number. A few examples are given below.

CH_3Cl	CH_3CH_2Br	$CH_3CH_2CH_2I$	$CH_3CCl_2CH_3$
Chloromethane	Bromoethane	1-Iodopropane	2,2-Dichloropropane

Halogenoalkanes are classified as **primary**, **secondary** and **tertiary**. In a primary halogenoalkane, the halogen is bonded to a carbon which is bonded to one alkyl group and two hydrogen atoms. In a secondary halogenoalkane, the halogen is attached to a carbon atom which is bonded to one hydrogen and two alkyl groups; and in a tertiary halogenoalkane, to a carbon which is bonded to three alkyl groups.

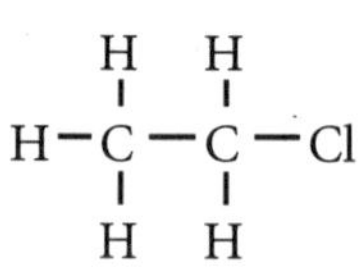

CH_3CH_2Cl

Chloroethane
(Primary)

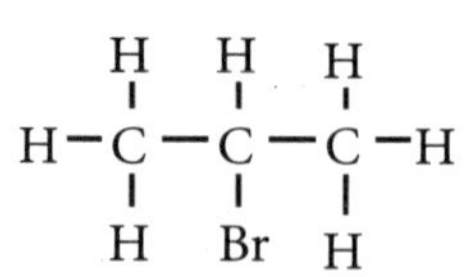

$CH_3CHBrCH_3$

2-Bromopropane
(Secondary)

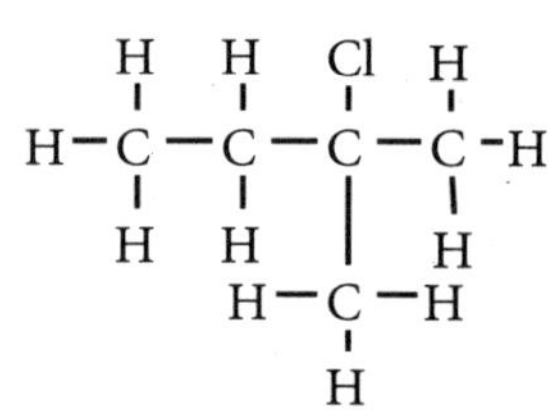

$CH_3CH_2CClCH_3$ (with CH_3 on the C bearing Cl)

2-Chloro-2-methylbutane
(Tertiary)

> **Practice question 8.4** Give the names of the following compounds. Classify them as primary, secondary or tertiary halogenoalkanes.

(a) CH_2ICH_3 (b) $CH_3CHBrCH_2CH_3$ (c) $CH_3\underset{CH_3}{\underset{|}{C}}ICH_3$

(d) $CH_3\overset{CH_3}{\overset{|}{\underset{CH_3}{\underset{|}{C}}}}CH_2Cl$

Reactions of halogenoalkanes

Halogenoalkanes are more reactive than alkanes due to the presence of a polar carbon-halogen bond. They undergo two important types of reactions: nucleophilic substitution and elimination.

Nucleophilic substitution

This is a reaction in which the halogen atom in a halogenoalkane is substituted by a nucleophile.

Note: Nucleophile (nucleus or positive liking) is a molecule or an anion which has a lone pair of electrons that can bond with another atom. Some examples of nucleophiles are H_2O, NH_3, ROH (alcohols), OH^- and Br^-.

When 2-bromopropane is treated with sodium hydroxide dissolved in a mixture of water and ethanol, 2-propanol is formed. In this reaction, the bromine atom is substituted by a hydroxyl group.

$$\underset{\text{2-Bromopropane}}{CH_3\underset{Br}{\underset{|}{C}}HCH_3} + OH^- \longrightarrow \underset{\text{2-Propanol}}{CH_3\underset{OH}{\underset{|}{C}}HCH_3} + Br^-$$

Elimination

When a halogenoalkane containing two or more carbon atoms in the molecule reacts with a base, the halogen atom and a hydrogen atom bonded to the next carbon are removed to form an alkene. This reaction is called elimination. Since a molecule of hydrogen halide is eliminated in this reaction, it is also called dehydrohalogenation. This is the reverse of the addition reaction of a hydrogen halide to an alkene.

When bromoethane is treated with a solution of potassium hydroxide in ethanol, a molecule of hydrogen bromide is eliminated from bromoethane to form ethene.

$$\underset{\text{Bromoethane}}{CH_3CH_2Br} \xrightarrow[-HBr]{\text{KOH/ethanol}} \underset{\text{Ethene}}{CH_2{=}CH_2}$$

8.7 Alcohols

Alcohols are compounds which contain a hydroxyl group (—OH) attached to an alkyl group. They can be considered as compounds formed by substituting a hydrogen atom of an alkane with a hydroxyl group. Alcohols form a homologous series with the general formula $C_nH_{2n+2}O$ or $C_nH_{2n+1}OH$, where *n* stands for the number of carbon atoms. The names of alcohols are derived from the corresponding alkanes by replacing the *-e* ending of alkane with *-ol*. The names and formulae of the first few members are given below.

Table 8.5

Alkane	Formula	Alcohol	Formula
Methane	CH_4	Methanol	CH_3OH
Ethane	C_2H_6	Ethanol	C_2H_5OH
Propane	C_3H_8	Propanol	C_3H_7OH
Butane	C_4H_{10}	Butanol	C_4H_9OH
Pentane	C_5H_{12}	Pentanol	$C_5H_{11}OH$

IUPAC rules to name alcohols

1. Find the longest C—C chain containing the hydroxyl group.
2. Number the carbon atoms in the chain, starting from the end closer to the hydroxyl group. The position of the hydroxyl group takes precedence over alkyl groups and halogens.
3. Note the location of the hydroxyl group by numbers and add the number as prefix or as shown in the examples.
4. Name the branches. Note the examples given below.

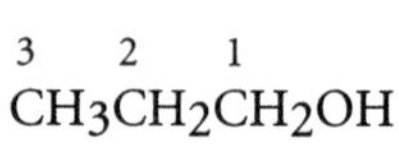

Propan-1-ol

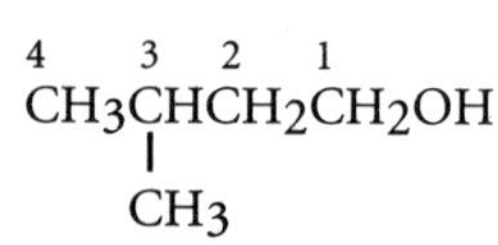

3-Methylbutan-1-ol

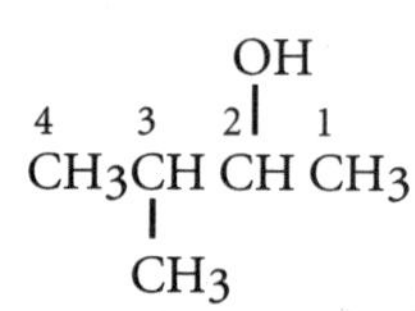

3-Methylbutan-2-ol

Alcohols are classified as **primary**, **secondary** and **tertiary** depending on the position of the hydroxyl group in the molecule. A primary alcohol has a —CH_2OH group i.e. the alcoholic carbon is bonded to at least two hydrogen atoms. A secondary alcohol has a —CHOH group and the secondary carbon is bonded to two alkyl groups and a hydrogen atom. A tertiary alcohol has a —COH group and the tertiary carbon is bonded to three alkyl groups.

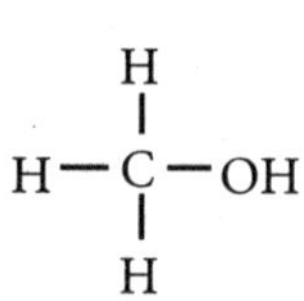

CH_3OH

Methanol
(Primary)

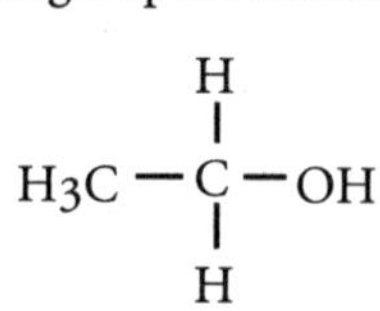

CH_3CH_2OH

Ethanol
(Primary)

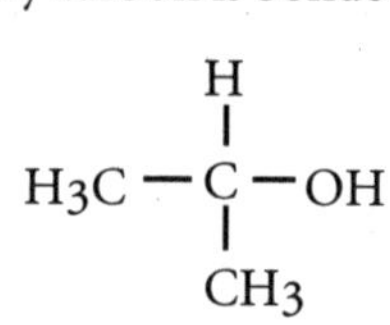

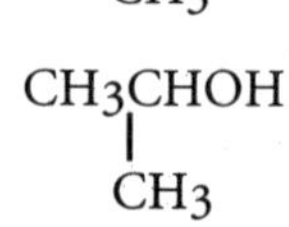

2-Propanol
(Secondary)

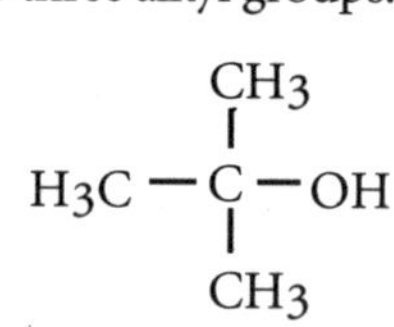

2-Methyl-2-propanol
(Tertiary)

Practice question 8.5 Write IUPAC names for these alcohols

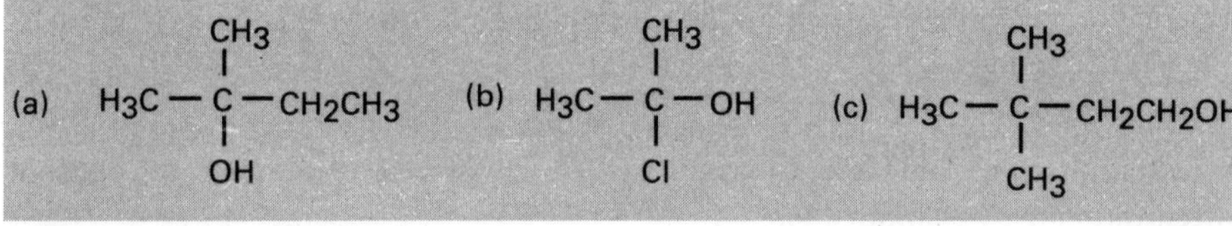

Properties of alcohols

Alcohols are colourless liquids and their melting and boiling points increase gradually as the molecular weight increases down the homologous series. The melting and boiling points of alcohols are generally higher than those of many organic compounds of similar molecular weights due to the strong intermolecular forces of attraction. The —OH group in alcohols is polar with a partial positive charge (δ+) on hydrogen and a partial negative charge (δ–) on oxygen. These hydroxyl groups form hydrogen bonds with other alcohol molecules.

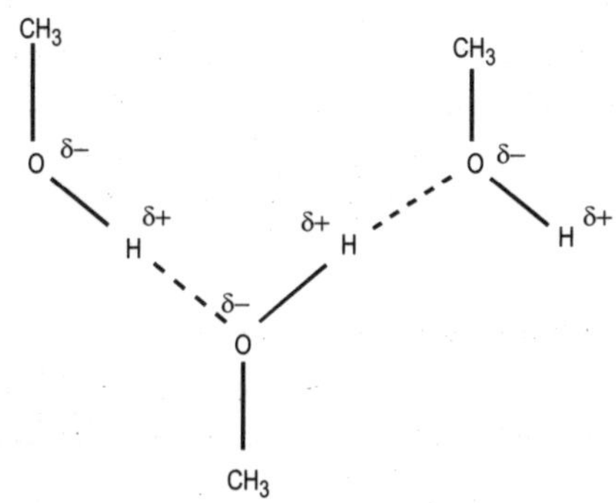

Hydrogen bonding between methanol molecules, shown by dotted lines

Figure 8.2

Lower alcohols (methanol, ethanol and propanol) are miscible with water, since water and alcohol molecules form intermolecular hydrogen bonds. Butanol is partially soluble. Solubility in water decreases as the size of the hydrocarbon section increases.

Ethanol is the most important alcohol. It is used in the form of alcoholic beverages, in the preparation of organic compounds and as an important organic solvent. Humans consume alcohol in the form of beer, wine and spirits. Beer contains about 2–5% ethanol, wine, 9 - 13% and whisky and brandy about 40%. Consuming alcohol in moderate amount gives a relaxed feeling while in large quantities it is harmful to health as it can affect the function of many organs, especially the liver.

Preparation of ethanol

Ethanol is prepared by the fermentation of fruit juices like grape juice. Fruit juice contains sugar, mainly glucose. When it is mixed with yeast at about 20° C, fermentation takes place. An enzyme, zymase, present in yeast acts as a catalyst to break down glucose to ethanol and carbon dioxide. The evolution of bubbles of carbon dioxide can be seen during fermentation. The yeast cells live on the energy evolved during the fermentation reaction, the equation for which is given below.

$$C_6H_{12}O_6(aq) \longrightarrow 2C_2H_5OH(aq) + 2CO_2(g)$$

Fermentation takes 2–3 days to complete. The yeast cells 'die' when the alcohol concentration reaches about 14%.

Spirits with high concentrations of alcohol are obtained by the distillation of the fermented liquid. Fractional distillation gives almost pure alcohol (99.4%) with traces of water.

Commercially, ethanol is also prepared from ethene. Ethene is a by-product of petroleum refining. Ethene is hydrated to form ethanol under suitable conditions. When ethene is treated with a mixture of dilute sulphuric acid and mercury (II) sulphate (as catalyst) under high pressure, the addition of water to ethene takes place to form ethanol.

$$CH_2{=}CH_2(g) + H_2O(l) \xrightarrow{H^+} CH_3CH_2OH$$

One main use of ethanol is to prepare *methylated spirit* which is ethanol adulterated with methanol. Methanol is poisonous and addition of it to ethanol makes it unfit for human consumption.

Reactions of alcohols

1. Reactive metals like sodium and potassium react with alcohols to give metal alkoxides and hydrogen. For example, ethanol reacts with sodium giving sodium ethoxide and hydrogen. Metal alkoxides (Example, sodium ethoxide) are ionic compounds.

$$2C_2H_5OH(l) + 2Na(s) \longrightarrow 2C_2H_5O^-Na^+(s) + H_2(g)$$

2. When alcohols react with hydrogen halides, phosphorus trichloride, phosphorus tribromide, phosphorus pentachloride or sulphonyl chloride, halogenoalkanes are formed. In these reactions, the hydroxyl group of the alcohol is substituted by the halogen atom. Note the examples given below and the other reaction products.

$$\underset{\text{Ethanol}}{CH_3CH_2OH} + HCl \longrightarrow \underset{\text{Chloroethane}}{CH_3CH_2Cl} + H_2O$$

$$\underset{\text{Ethanol}}{3CH_3CH_2OH} + PBr_3 \longrightarrow \underset{\text{Bromoethane}}{3CH_3CH_2Br} + H_3PO_3$$

$$\underset{\text{1-Propanol}}{CH_3CH_2CH_2\,OH} + PCl_5 \longrightarrow \underset{\text{1-Chloroethane}}{CH_3CH_2CH_2\,Cl} + POCl_3 + HCl$$

$$\underset{\text{Ethanol}}{CH_3CH_2OH} + SOCl_2 \longrightarrow \underset{\text{Chloroethane}}{CH_3CH_2Cl} + SO_2 + HCl$$

3. **Dehydration**

When an alcohol is heated with a dehydrating agent such as concentrated sulphuric acid or 85% phosphoric acid, dehydration (removal of a molecule of water) takes place and an alkene is formed. The hydroxyl group and a hydrogen bonded to the next carbon are removed as a molecule of water. Ethanol on dehydration gives ethene and propanol gives propene.

$$\underset{\text{Ethanol}}{CH_3CH_2OH} \xrightarrow[-H_2O]{H_2SO_4} \underset{\text{Ethene}}{CH_2{=}CH_2}$$

$$\underset{\text{Propan-1-ol}}{CH_3CH_2CH_2OH} \xrightarrow[-H_2O]{H_2SO_4} \underset{\text{Propene}}{CH_2CH = CH_2}$$

4. **Oxidation of alcohols** An oxidising agent oxidises a primary alcohol to an aldehyde, which is further oxidised to a carboxylic acid. Oxidising agents commonly used are acidified potassium permanganate ($KMnO_4$), acidified potassium dichromate ($K_2Cr_2O_7$) and chromium (VI) oxide (CrO_3). A secondary alcohol is oxidised to a ketone and a tertiary alcohol is not oxidised without degrading the molecule.

$$\underset{\substack{\text{Ethanol}\\ \text{(A primary alcohol)}}}{CH_3CH_2OH} \xrightarrow[H_2SO_4]{CrO_3} \underset{\substack{\text{Ethanal}\\ \text{(An aldehyde)}}}{CH_3CHO} \xrightarrow[H_2SO_4]{CrO_3} \underset{\substack{\text{Ethanoic acid}\\ \text{(A carboxylic acid)}}}{CH_3COOH}$$

$$\underset{\substack{\text{2-Propanol}\\ \text{(A secondary alcohol)}}}{CH_3CHOHCH_3} \xrightarrow[H_2SO_4]{CrO_3} \underset{\substack{\text{Propanone}\\ \text{(A ketone)}}}{CH_3COCH_3}$$

5. **Combustion** Alcohols undergo combustion reactions exothermically, producing carbon dioxide and water.

$$C_2H_5OH(l) + 3O_2(g) \longrightarrow 2CO_2(g) + 3H_2O(l)$$

8.8 Summary of reaction types in this chapter

Combustion of an organic substance is a chemical reaction in which the substance combines with oxygen to give carbon dioxide, water and heat.

Substitution (replacement) is a reaction in which an atom (or group) is substituted by another atom (group).

Addition is a reaction in which a molecule is added to another molecule to form a single product.

Elimination is a reaction in which small molecules like hydrogen halides (HX) or water are removed from a molecule. For example, the elimination of hydrogen and halogen from the adjacent carbon atoms of a halogenoalkane or hydrogen and hydroxyl of an alcohol, gives an alkene.

Hydration is addition of the water to a substance.

Hydrolysis is the breakdown of a substance by its reaction with water.

Polymerisation is a reaction in which many small molecules called monomers bond together to form large molecules called polymers.

Tutorial: helping you learn

Progress questions

8.1 Draw the Lewis structure (electron dot formula) of the following compounds.
(a) CH_3Cl (b) $CH_3CH{=}CH_2$ (c) CH_3OH

8.2 Write the elaborate and condensed structural formulae of each of the following.
(a) 2, 2-dimethylbutane (b) 1,4-dichlorobut-2-ene (c) 2-methylpropan-2-ol
(d) 2-bromopropan-1-ol (e) pent-2-ene (f) 2,3-dimethylbut-1-ene

8.3 Give the IUPAC names of the following compounds.

(a) 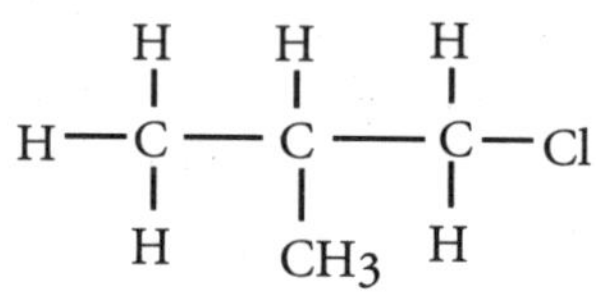

(b) 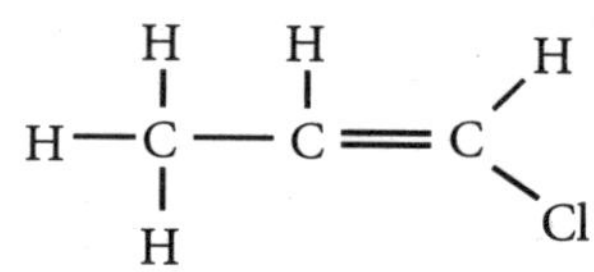

(c) 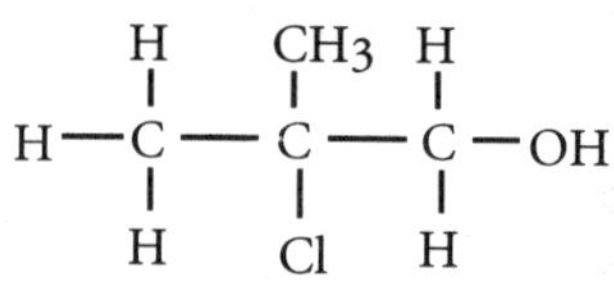

(d) 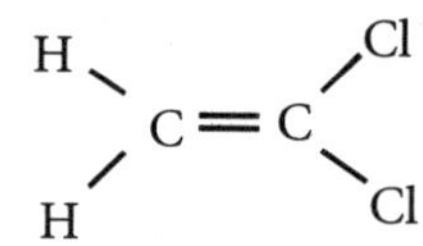

8.4 Draw the structure of these compounds. Give the name and draw the structure of an isomer of each of these compounds.
(a) 2-chloropropane (b) 1,2-dibromopropane
(c) 2-methylpropene (d) butan-2-ol

8.5 Give an explanation for each of the following observations.
(a) Ethanol is miscible with water while ethane is not.
(b) When ethene is passed through bromine water, decolourisation of bromine water takes place.
(c) Effervescence is seen when a piece of sodium metal is added to ethanol.

8.6 Name the main reaction products, and write an equation, when equimolar (equal number of moles) quantities of methane and bromine are allowed to react under suitable conditions.

8.7 Write balanced equations for the combustion of the following compounds.
(a) propane (b) propene (c) propanol

8.8 Write an equation for the reaction between propene and bromine water. Name the product.

8.9 Name the following reactions.

(a) $CH_3C \equiv CH + HCl \longrightarrow CH_3CCl = CH_2$
Propyne

(b) $2n \left[CH_3CCl = CH_2 \right] \longrightarrow$ 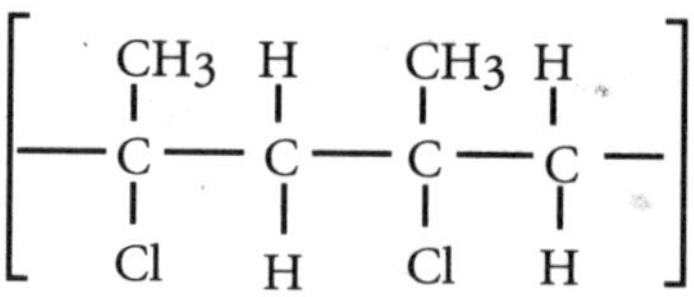
2-Chloropropene

9 Introduction to Inorganic Chemistry

One-minute overview

Inorganic chemistry is the study of the nature and properties of inorganic substances, that is, all elements and their compounds other than carbon compounds. There are only a few elements in nature in the uncombined form most elements being present in the form of compounds. In this chapter we shall discuss the extraction of some metals from their compounds and the gradation of properties of Group I, Group II and Group VII elements and their compounds, also how elements of the same group have similar properties because of the similarity in their outer electron shell configuration.

Topics discussed in this chapter include:

- purification of bauxite
- electrolytic extraction of aluminium
- extraction of iron in a blast furnace
- preparation and reduction of titanium (IV) chloride
- Group I elements; alkali metals
- Group II elements; alkaline earth metals
- Group VII elements; halogens
- hydrogen halides

9.1 Extraction of metals

We have discussed metallic bonding in Chapter 3 (Section 3.7). Metals vary in reactivity from the very unreactive metals (*Example*, gold and platinum) to the very reactive ones (*Example*, sodium and potassium). Metals are arranged in their order of reactivity using electrode potential values which shall be explained in another section. This arrangement of metals in the order of reactivity is called the **activity series**. Below is outlined some common metals in the order of reactivity.

K
Ca
Na
Mg
Al
Zn
Fe
Sn
Pb
H
Cu
Ag
Hg

↑ Reactivity increases

Figure 9.1: Activity series

Note: Though hydrogen is a non-metal, it is placed together with metals in the activity series, since it is an electropositive element and it can donate its electron during chemical reactions to become a positively charged ion, like metals.

We have seen in Chapter 7 that metals are reducing agents and they release electrons during chemical reactions. Reactive metals lose electrons more readily than less reactive metals. For example, calcium which is above aluminium in the activity series is more reactive than aluminium and calcium atom loses electrons more easily than aluminium does.

$$Ca \longrightarrow Ca^{2+} + 2e$$
$$Al \longrightarrow Al^{3+} + 3e$$

Most metals are present in the earth's crust in the form of compounds. Metals are extracted from these compounds. In a compound, the metal is present as positively charged ions. The extraction of the metal involves the reduction of the positively charged ions to the metal. This is the reverse of the oxidation reaction mentioned above.

$$Ca^{2+} + 2e \longrightarrow Ca$$
$$Al^{3+} + 3e \longrightarrow Al$$

Calcium ions accept electrons to form calcium, and aluminium ions similarly form aluminium. These are reduction reactions. It is more difficult to reduce calcium ions than to reduce aluminium ions. As the reactivity of the metal in the series decreases, the reactivity of its metal ion increases.

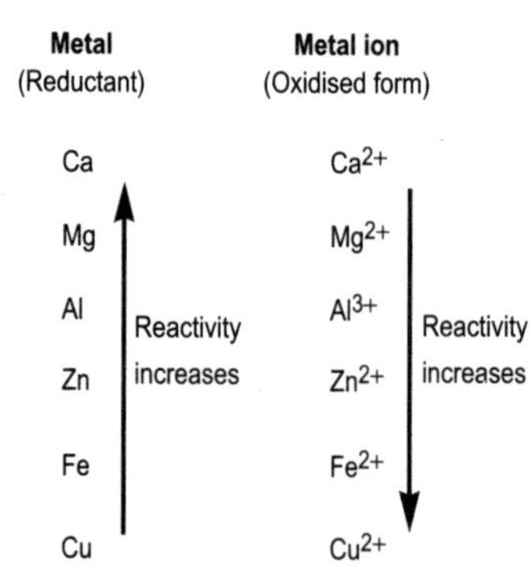

Figure 9.2

Note: A metal can displace a less reactive metal from its compound in aqueous solution. For example, magnesium can displace zinc from zinc sulphate solution or zinc can displace copper from copper (II) sulphate solution. In these cases the more reactive metal gives out electron to the ion of the less reactive metal. These are redox reactions.

$$Mg(s) + ZnSO_4(aq) \longrightarrow MgSO_4(aq) + Zn(s)$$
$$Zn(s) + CuSO_4(aq) \longrightarrow ZnSO_4(aq) + Cu(s)$$

Example 9.1 Magnesium displaces zinc from zinc sulphate solution. Write partial equations for the oxidation of Mg and for the reduction of Zn^{2+} ion. Combine the two equations to get the overall equation for the redox reaction.

Solution

$$Mg \longrightarrow Mg^{2+} + 2e^-$$
$$Zn^{2+} + 2e^- \longrightarrow Zn$$
$$\overline{Mg + Zn^{2+} \longrightarrow Mg^{2+} + Zn}$$

Only very few metals like gold and platinum are found in the earth's crust in the uncombined form. A **mineral** is a naturally occurring inorganic compound. An **ore** is a mineral from which a metal is extracted.

The preparation of a metal from an ore involves a number of steps. First the ore is collected and concentrated or purified in order to remove sand, clay etc. The next step is to reduce the ore to the metal by a suitable reduction process or by using a suitable reducing agent. Finally the metal is purified if necessary.

The type of the reduction method in converting ore to metal depends mainly on the reactivity of the metal. The compounds of the reactive metals at the top of the activity series can not be reduced to metals by common reducing agents like hydrogen, carbon or carbon monoxide. Such metals are obtained by the electrolytic reduction of their compounds in the molten state (*Example*, sodium and aluminium). The next set of metals are obtained by the reduction of their oxides by carbon or carbon monoxide (*Example*, iron and lead).

Aluminium

Aluminium is extracted from the ore, **bauxite**, which is hydrated aluminium oxide ($Al_2O_3 \cdot xH_2O$). The ore is usually contaminated with two main impurities; silica (sand, SiO_2) and iron (III) oxide (Fe_2O_3). First of all, the ore is purified in order to get pure aluminium oxide and this is done by a chemical process. Then, aluminium metal is obtained from the purified aluminium oxide by an electrolytic method.

Purification of bauxite

The crushed ore is mixed with hot concentrated sodium hydroxide solution. Aluminium oxide and silica dissolve in the alkali to form sodium aluminate and sodium silicate respectively, while iron oxide remains insoluble and is removed by filtration.

$$Al_2O_3(s) + 2NaOH(aq) + 3H_2O(l) \longrightarrow \underset{\text{sodium aluminate}}{2NaAl(OH)_4(aq)}$$

$$SiO_2(s) + 2NaOH(aq) \longrightarrow \underset{\text{sodium silicate}}{Na_2SiO_3(aq)} + H_2O(l)$$

$$Fe_2O_3(s) - \text{insoluble in } NaOH(aq)$$

The solution containing sodium aluminate and silicate is heated under high pressure with some freshly prepared aluminium hydroxide solid as 'seed.' Sodium aluminate is converted to aluminium hydroxide which grows on the seeded aluminium hydroxide. Sodium silicate does not hydrolyse and remains in solution in the soluble form.

$$NaAl(OH)_4(aq) \xrightarrow{seed} NaOH(aq) + Al(OH)_3(s)$$

Aluminium hydroxide is filtered and heated at about 1000° C when it decomposes to give pure aluminium oxide.

$$2Al(OH)_3(s) \longrightarrow Al_2O_3(s) + 3H_2O(g)$$

Aluminium oxide is electrolysed to get aluminium.

Electrolysis of aluminium oxide

For the electrolysis, a large steel tank lined with graphite is used. The graphite lining acts as the cathode. The anode is made up of blocks of graphite which are dipped in the molten electrolyte. Aluminium oxide has a very high melting point (2050° C). It would need a large amount of energy to keep it in the molten state for the electrolysis. But aluminium oxide is soluble in molten cryolite (sodium hexafluoroaluminate, Na_3AlF_6) and the mixture can be kept in the molten state at about 850° C for the electrolysis.

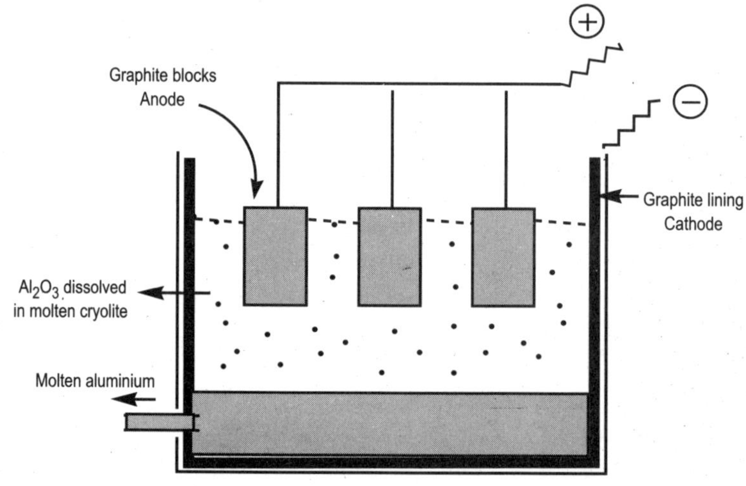

Figure 9.3: Electrolytic extraction of aluminium

During the electrolysis, positive Na^+ and Al^{3+} ions are attracted to the cathode. Al^{3+} ions accept electrons in preference to Na^+ ions since aluminium is below sodium in the activity series. Each Al^{3+} ion takes three electrons and is reduced to an Al atom.

At the cathode,

$$Al^{3+} + 3e^- \longrightarrow Al$$

Thus, aluminium metal is formed at the cathode lining in the molten state because of the high temperature of the bath. Since it is heavier than the electrolyte, it is collected at the bottom and can be drawn out periodically.

The anions O^{2-} and AlF_6^- are attracted to the anode. There, O^{2-} ions lose electrons in preference to AlF_6^- ions. Each O^{2-} ion loses two electrons to form an O atom and these oxygen atoms combine with the carbon anode to form carbon dioxide.

At the anode,

$$O^{2-} \longrightarrow O + 2e$$
$$2[O] + C \longrightarrow CO_2(g)$$

Since the anode blocks react during electrolysis and are eaten away, they are replaced periodically.

Aluminium is a strong metal. It has a low density compared to many metals and is a good conductor of electricity. Though it is a very reactive metal, it is resistant to corrosion because aluminium forms a thin invisible oxide film on the surface of the metal by atmospheric oxidation. This oxide layer protects aluminium from corrosion. These qualities make aluminium a very useful metal for making overhead powerlines, aeroplanes and other architectural constructions.

Iron

Two important ores of iron are **haematite,** Fe_2O_3 and **magnetite,** Fe_3O_4. Iron is extracted from these ores by a reduction process in a large furnace (*blast furnace*).

A blast furnace is about 80–100 feet high and is lined inside with heat resistant bricks. The ore (8 parts) which is usually contaminated with silica, is crushed, mixed with coke, C (4 parts) and lime stone, $CaCO_3$ (1 part) and is fed into the top of the furnace.

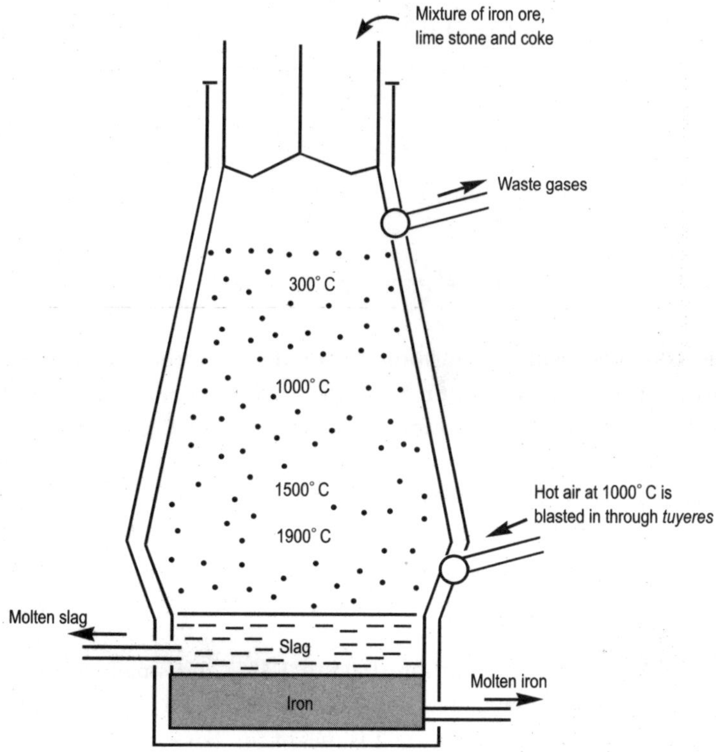

Figure 9.4: Blast furnace

A blast of hot air at about 1000° C is forced into the furnace through pipes known as *tuyeres* (See the diagram). The coke in the mixture burns exothermically to produce carbon dioxide and the temperature of this region rises to about 1900° C.

$$C(s) + O_2(g) \longrightarrow CO_2(g), \quad \Delta H = -393.5 \text{ kJ mol}^{-1}$$

As the hot gases in the furnace rise through the 'charge', carbon dioxide combines with the carbon (coke) to form carbon monoxide. This is an endothermic reaction and the temperature of this region is lower than the one below (See the diagram).

$$CO_2(g) + C(s) \longrightarrow 2CO(g), \quad \Delta H = +172.0 \text{ kJ mol}^{-1}$$

Carbon monoxide reduces the iron oxides to iron.

$$Fe_2O_3(s) + 3CO(g) \longrightarrow 2Fe(s) + 3CO_2(g)$$
$$Fe_3O_4(s) + 4CO(g) \longrightarrow 3Fe(s) + 4CO_2(g)$$

As the iron descends, it melts in the hot region of the furnace, dissolving some carbon and is collected at the bottom of the furnace. The molten iron is drawn out periodically into moulds of sand and cooled. This iron is known as **pig iron.**

The limestone decomposes to calcium oxide and carbon dioxide. Carbon dioxide joins the rest of the gases. Calcium oxide acts as a *flux* and combines with the silica *impurity* to form *slag*, calcium silicate. It also melts and is collected on the top of the molten iron as it is less dense than molten iron. This is taken out through the slag hole, cooled and used for the construction of buildings and for road surfacing.

$$CaCO_3(s) \longrightarrow CaO(s) + CO_2(g)$$
$$CaO(s) + SiO_2(s) \longrightarrow CaSiO_3(s)$$

Note: A compound such as calcium oxide which combines with and lowers the melting point of an impurity in an ore to form a slag, is called flux. A flux combines with an impurity to form a slag.

Flux + impurity ⟶ slag

Pig iron contains many impurities; about 4–5% carbon and some silicon, phosphorus, manganese and sulphur. Pig iron is brittle because of the high carbon content and has a relatively low melting point (about 1200° C). Pig iron is converted to **cast iron** by heating it with scrap iron, steel and coke. Cast iron contains about 2% carbon and can be cast into various forms.

Wrought iron is about 99.5% pure iron, and has a melting point of about 1400° C. It is prepared from pig iron by heating it with haematite and lime stone in a furnace to remove most of the impurities. Wrought iron is strong, malleable and ductile.

Steel is an alloy of iron with carbon (0.03 to 1.4%) and small amounts of other metals such as chromium, manganese etc. which are added to impart special desired properties depending on the type of use it is meant for. For example, stainless steel is highly corrosion resistant and contains up to 15% chromium. The tensile strength, hardness and malleability of steel varies with its carbon content. Higher carbon content makes the iron harder and more brittle.

One method to prepare steel is by a process called the **basic oxygen process.** Oxygen and powdered limestone are forced through molten pig iron when most of the carbon burns, and lime stone reacts with the impurities to form molten slag. The slag is removed and appropriate amounts of metals are added to the molten iron to form the required steel.

Titanium

Two common ores of titanium are **rutile** (TiO_2) and **ilmenite** ($FeO \cdot TiO_2$). Titanium metal is extracted from titanium (IV) oxide, TiO_2, by the **Kroll process.** First, the oxide is heated with carbon in a stream of chlorine gas at about 800–1000° C, when titanium (IV) oxide is converted to titanium (IV) chloride.

$$TiO_2(s) + C(s) + 2Cl_2(g) \longrightarrow TiCl_4(g) + CO_2(g)$$

Titanium (IV) chloride is a liquid (B.pt. 136° C) and can be purified by fractional distillation.

In the next step of extraction, titanium (IV) chloride is heated with molten magnesium metal in an atmosphere of argon in a steel furnace lined with molybdenum. Magnesium reduces titanium (IV) chloride to titanium. Magnesium chloride is formed simultaneously.

$$TiCl_4(l) + 2Mg(l) \longrightarrow Ti(s) + 2MgCl_2(l)$$

The molten magnesium chloride is separated from 'titanium sponge' and is further purified by heating to remove any magnesium chloride impurity.

Magnesium chloride is converted to magnesium by the electrolysis of molten magnesium chloride and is recycled.

Titanium is a light and strong metal and is used extensively in the aircraft industry. Titanium (IV) oxide (titanium dioxide, TiO_2) is brilliantly white and non-toxic and is used in paint making and in the paper industry.

9.2 Group I elements: similarities and trends in properties

Group I elements – lithium, sodium, potassium, rubidium, caesium and francium are also known as *alkali metals.* Francium is a radioactive element and in properties resembles caesium. Each Period starts with an alkali metal except Period 1. Alkali metals have one electron in the outermost orbit of the atom. They have an outer shell configuration of s^1 (Table 9.1).

Alkali metals have low melting and boiling points compared to other metals. They are soft metals and can be cut with a knife.

Alkali metals have low first ionisation energies and so they are very reactive. As ionisation energy decreases down the group, reactivity increases from lithium to caesium. They react with non-metals to form ionic compounds. The alkali metal atom donates its valence electron during the reaction and form a stable +1 metal ion. They are good reducing agents.

Table 9.1: Properties of Group I metals

	Li	Na	K	Rb	Cs
Atomic number	3	11	19	37	55
Outer shell configuration	$2s^1$	$3s^1$	$4s^1$	$5s^1$	$6s^1$
Melting point (°C)	181	98	63	39	29
1st ionization energy (kJ mol^{-1})	520	496	419	403	376
Electronegativity	1.0	0.9	0.8	0.8	0.7

The second ionisation energies of alkali metals are very high as these correspond to the removal of an electron from a noble gas configuration. So alkali metals do not form compounds in the +2 oxidation state.

Practice question 9.1

(a) Write the electronic configurations of Na ($Z = 11$) and Na^+ ion.

(b) The first ionisation energy of Na is 496 kJ mol^{-1} and its second ionization energy is 4563 kJ mol^{-1}. Give an explanation for the large difference in these values.

(c) Do you expect sodium to form Na^{2+} ions during chemical reactions? Why?

The relative reactivity of Group I elements and their compounds

Reaction with oxygen

Alkali metals react with oxygen readily. They tarnish in air because of the reaction with oxygen and moisture, and formation of a coating of the products. So they are stored in liquid paraffin. Lithium reacts with oxygen to form lithium oxide.

$$4Li(s) + O_2(g) \longrightarrow 2Li_2O(s)$$

Sodium reacts with oxygen to give a mixture of sodium oxide and sodium peroxide.

$$4Na(s) + O_2(g) \longrightarrow 2Na_2O(s)$$
$$2Na(s) + O_2(g) \longrightarrow Na_2O_2(s)$$

Potassium with oxygen gives a mixture of potassium oxide, peroxide and superoxide.

$$4K(s) + O_2(g) \longrightarrow 2K_2O(s)$$
$$2K(s) + O_2(g) \longrightarrow \underset{\text{Potassium peroxide}}{K_2O_2(s)}$$
$$K(s) + O_2(g) \longrightarrow \underset{\text{Potassium superoxide}}{KO_2(s)}$$

Reaction with water

Alkali metals react with water to give metal hydroxides and hydrogen. Lithium reacts slowly, while other metals react vigorously, the intensity of the reaction increasing from sodium to caesium. The metal hydroxides are soluble in water and form strong alkalis.

$$2Na(s) + 2H_2O \longrightarrow 2NaOH(aq) + H_2(g)$$

Reaction with non-metals

Alkali metals react with a number of non-metals to form ionic compounds. For example, they react with halogens (F_2, Cl_2, Br_2 and I_2) to form metal halides, with sulphur to form metal sulphides and with hydrogen to form metal hydrides.

$$2Na(s) + Cl_2(g) \longrightarrow 2NaCl(s)$$

$$2Na(s) + S(s) \longrightarrow \underset{\text{Sodium sulphide}}{Na_2S(s)}$$

$$2Na(s) + H_2(g) \longrightarrow \underset{\text{Sodium hydride}}{2NaH(s)}$$

Example 9.2 Write the formulae of the ions present in these compounds

(a) Potassium fluoride, KF (b) sodium oxide, Na_2O

(c) potassium sulphide, K_2S (d) lithium hydroxide, LiOH

Solution

(a) K^+ and F^- (b) Na^+ and O^{2-} (c) K^+ and S^{2-} (d) Li^+ and OH^-

Solubility and stability of compounds of Group I elements

Alkali metal compounds are generally soluble in water. Thus, oxides, hydroxides, chlorides, nitrates, sulphates and carbonates are all soluble in water.

An alkali metal oxide dissolves in water to give a solution of the metal hydroxide.

$$Na_2O(s) + H_2O(l) \longrightarrow 2NaOH(aq)$$

Alkali metal oxides, hydroxides and carbonates are stable and are not decomposed by heat except lithium hydroxide and lithium carbonate, which decompose on heating to give lithium oxide.

$$2LiOH(s) \longrightarrow Li_2O(s) + H_2O(l)$$

$$Li_2CO_3(s) \longrightarrow Li_2O(s) + CO_2(g)$$

Alkali metal nitrates decompose on heating to produce nitrites and oxygen, except lithium nitrate which gives rise to the oxide, nitrogen dioxide and oxygen.

$$4LiNO_3(s) \longrightarrow 2Li_2O(s) + 4NO_2(g) + O_2(g)$$

$$\underset{\text{Sodium nitrate}}{2NaNO_3(s)} \longrightarrow \underset{\text{Sodium nitrite}}{2NaNO_2(s)} + O_2(g)$$

Practice question 9.2 Write equations for the reaction between (a) potassium and water, (b) potassium oxide and water, (c) potassium and bromine, and (d) potassium and sulphur.

9.3 Group II elements: similarities and trends in properties

The Group II elements – beryllium, magnesium, calcium, strontium, barium and radium are metals except beryllium which has more non-metallic than metallic properties. They are called *alkaline earth metals*. They all contain two electrons in the outermost energy level with ns^2 electronic configuration. Radium is a radioactive element and its properties are similar to those of barium. Some of the physical characteristics of the Group II elements are summarised in Table 9.2.

Group II metals have higher melting and boiling points than the corresponding Group I metals. This is because of the stronger metallic bonding that exists between atoms. There are two valence electrons on each atom which contribute to the delocalised electron cloud and the atomic kernels are doubly positively charged (Section 3.10). Therefore there is stronger attraction between atomic kernels and the delocalised electron cloud.

Table 9.2: Properties of Group II metals

	Be	Mg	Ca	Sr	Ba
Atomic number	4	12	20	38	56
Outer shell configuration	$2s^2$	$3s^2$	$4s^2$	$5s^2$	$6s^2$
Melting point (°C)	1280	650	838	770	714
Atomic radius (nm)	0.112	0.160	0.197	0.215	0.222
Ionic radius (nm)	0.031	0.065	0.099	0.113	0.135
1st ionization energy (kJ mol^{-1})	900	738	590	550	503
2nd ionization energy (kJ mol^{-1})	1757	1451	1145	1064	965
Electronegativity	1.5	1.2	1.0	1.0	0.9

The first ionisation energy of a Group II metal is higher than that of the corresponding Group I metal (Section 3.1). The second ionisation energy is higher than the first ionisation energy, and not as high as the second ionisation energy of the corresponding Group I metal.

Group II metal atoms donate two electrons during chemical reactions to form +2 positive metal ions, with a stable noble gas configuration of eight electrons in the outer shell. So all Group II metals have a +2 oxidation number and they form ionic compounds. Group II metals are also very reactive, though not as reactive as the corresponding Group I metals and the reactivity increases down the group.

The relative reactivity of Group II elements and their compounds

Reaction with oxygen

Group II metals react with oxygen to form oxides. For example, beryllium on heating in oxygen produces beryllium oxide. Magnesium burns in oxygen with a brilliant white flame to form magnesium oxide.

$$2Be(s) + O_2(g) \longrightarrow 2BeO(s)$$
$$2Mg(s) + O_2(g) \longrightarrow 2MgO(s)$$

Reaction with nitrogen

Group II metals react with nitrogen on heating to form metal nitrides. For example, magnesium gives magnesium nitride.

$$3Mg(s) + N_2(g) \longrightarrow Mg_3N_2(s)$$

Reaction with halogens

Group II metals react with halogens to give metal halides. Example,

$$Ca(s) + Br_2(g) \longrightarrow CaBr_2(s)$$

Reaction with water

Beryllium and magnesium react slowly with water. Magnesium reacts with steam more readily to produce magnesium oxide and hydrogen.

$$Mg(s) + H_2O(g) \longrightarrow MgO(s) + H_2(g)$$

Calcium reacts with water at room temperature, with effervescence and evolution of hydrogen gas, to form a suspension of calcium hydroxide. Strontium and barium react in a similar manner, but more vigorously.

$$Ca(s) + 2H_2O(l) \longrightarrow Ca(OH)_2(aq) + H_2(g)$$

Practice question 9.3 Calcium combines with bromine to form a compound of formula $CaBr_2$ and not CaBr. Why?

Solubility and stability of Group II metal compounds

Group II metal chlorides and nitrates are soluble in water. Beryllium and magnesium sulphates are soluble in water while the sulphates of the other Group II metals are insoluble and solubility decreases down the series.

$$BeSO_4 > MgSO_4 > CaSO_4 > SrSO_4 > BaSO_4$$
$$\xrightarrow{\text{Solubility decreases}}$$

The hydroxides of beryllium and magnesium are insoluble in water, calcium hydroxide is partially soluble and the hydroxides of strontium and barium more soluble. Solubility increases down the series.

$$Be(OH)_2 < Mg(OH)_2 < Ca(OH)_2 < Sr(OH)_2 < Ba(OH)_2$$
$$\xrightarrow{\text{Solubility increases}}$$

The carbonates of Group II metals decompose on heating to give the oxides and carbon dioxide. The ease of decomposition decreases down the series.

$$MgCO_3(s) \longrightarrow MgO(s) + CO_2(g)$$

The nitrates of Group II metals decompose on heating to produce the oxides, nitrogen dioxide and oxygen.

$$2Ca(NO_3)_2(s) \longrightarrow 2CaO(s) + 4NO_2(g) + O_2(g)$$

9.4 Group VII elements: similarities and trends in properties

Group VII elements – fluorine, chlorine, bromine and iodine are called *halogens.* Fluorine and chlorine are gases, bromine, a liquid and iodine, a solid. Bromine vapour is brown and iodine vapour, purple. They all form diatomic covalent molecules of general formula X_2, where X stands for the halogen atom. There is close resemblance in their chemical properties and a gradation in physical properties. Some properties of halogens are summarised in Table 9.3.

Halogen atoms contain seven electrons in the outermost energy level with an outer shell electronic configuration of ns^2np^5. They have the tendency to gain one electron in order to attain the stable electronic configuration of noble gases.

Halogens are non-metals. They have fairly low boiling points as the intermolecular forces of attraction between X_2 molecules are van der Waals forces (Table 9.3). The boiling point increases down the group because the strength of van der Waals forces increases as molecular weight increases down the series. Atomic and ionic radii increase down the group as expected, since the number of energy levels increases. (Compare the atomic and ionic radii of halogens in Table 9.3.)

Table 9.3: Properties of halogens

	F	Cl	Br	I
Atomic number	9	17	35	53
Outer shell electronic configuration	$2s^22p^5$	$3s^23p^5$	$4s^24p^5$	$5s^25p^5$
Physical appearance	Pale yellow gas	Greenish yellow gas	Brown liquid	Black shiny solid
Boiling point (°C)	–187	–35	59	183
Atomic radius (nm)	0.071	0.099	0.114	0.133
Ionic radius (nm)	0.136	0.181	0.195	0.216
Ionization energy (kJ mol^{-1})	1681	1251	1140	1008
Electronegativity	4.0	3.0	2.8	2.5
Bond energy (kJ mol^{-1})	158.0	243.4	193.0	151.2

The first ionisation energies of halogens are generally high. They do not lose electrons easily. Ionisation energy decreases down the series, as the size of the atom increases, and as the screening effect increases. Halogens have high values of electronegativity. Fluorine is the most electronegative element (Section 3.6).

The relative reactivity of Group VII elements and their compounds

Group VII elements are very reactive. They are strong oxidising agents and accept electrons to form halide (X^-) ions. They also share electrons with other atoms to form covalent compounds. Fluorine is the most reactive halogen and reactivity decreases down the group. They react with a number of substances including water, hydrogen, metals, organic compounds and alkalis.

Reaction with hydrogen

Halogens react with hydrogen to form hydrogen halides. Fluorine and chlorine react with explosive violence and bromine and iodine react on heating.

$$H_2(g) + Cl_2(g) \longrightarrow 2HCl(g)$$

Reaction with water

Fluorine reacts with water to produce hydrofluoric acid and oxygen gas.

$$2F_2(g) + 2H_2O(l) \longrightarrow 4HF(aq) + O_2(g)$$

Chlorine dissolves in water to form 'chlorine water' which is a mixture of hydrochloric acid and hypochlorous acid. Bromine reacts in a similar way.

$$Cl_2(g) + H_2O(l) \longrightarrow HCl(aq) + HOCl(aq)$$

Iodine is only slightly soluble in water, but dissolves well in a solution containing iodide ion, I^-, because I_2 combines with I^- to form a soluble complex ion, I_3^-.

$$I_2(s) + KI(aq) \longrightarrow KI_3(aq)$$

$$\text{Or,} \quad I_2(s) + I^-(aq) \longrightarrow I_3^-(aq)$$

Reaction with metals

Halogens react with most metals to form metal halides. A few examples are shown below.

$$2Na(s) + F_2(g) \longrightarrow 2NaF(s)$$

$$Ca(s) + Cl_2(g) \longrightarrow CaCl_2(s)$$

Reaction with alkalis

Fluorine reacts with sodium or potassium hydroxide solution to produce oxygen difluoride.

$$2F_2(g) + 2NaOH(aq) \longrightarrow 2NaF(aq) + H_2O(l) + OF_2(g)$$

Chlorine reacts with dilute sodium hydroxide solution to give sodium chloride, sodium chlorate (I), NaClO, (sodium hypochlorite) and water. Bromine and iodine react in a similar manner.

$$Cl_2(g) + 2NaOH(aq) \longrightarrow NaCl(aq) + NaClO(aq) + H_2O(l)$$

Chlorine reacts with hot, concentrated sodium hydroxide solution to give sodium chloride, sodium chlorate (V) and water. Bromine and iodine react with hot alkali in the same way.

$$3Cl_2(g) + 6NaOH(aq) \longrightarrow 5NaCl(aq) + NaClO_3(aq) + 3H_2O(l)$$

Reaction with halide ions

A halogen can displace another halogen below it in the series from its halide solution. For example, chlorine can displace bromine from a bromide solution or iodine from an iodide solution. Similarly, bromine can displace iodine from an iodide solution.

$$Cl_2(g) + 2KBr(aq) \longrightarrow 2KCl(aq) + Br_2(aq)$$
$$Cl_2(g) + 2Br^-(aq) \longrightarrow 2Cl^-(aq) + Br_2(aq) \text{ – ionic equation}$$
$$Br_2(aq) + 2KI(aq) \longrightarrow 2KBr(aq) + I_2(aq)$$

Practice question 9.4 A redox reaction takes place between chlorine and sodium hydroxide solution. During the reaction, Cl_2 undergoes oxidation as well as reduction. Write partial equations for the reduction of Cl_2 to Cl^- and oxidation of Cl_2 to ClO^- in the presence of OH^- ion. Write the overall equation.

9.5 Hydrogen halides

Hydrogen halides can be prepared in the laboratory by the reaction of metal halides with acids. Thus hydrogen fluoride is prepared by treating calcium fluoride with concentrated sulphuric acid, and hydrogen chloride by the reaction of sodium chloride with concentrated sulphuric acid.

$$CaF_2(s) + H_2SO_4(aq) \longrightarrow CaSO_4(s) + 2HF(g)$$
$$NaCl(s) + H_2SO_4(aq) \longrightarrow NaHSO_4(aq) + HCl(g)$$

Hydrogen bromide and hydrogen iodide are not prepared by the above method. This is because hydrogen bromide and hydrogen iodide are good reducing agents and they react with concentrated sulphuric acid, an oxidising agent, reducing it to sulphur dioxide while the hydrogen halides themselves are oxidised to the corresponding halogens. For example, sodium bromide with concentrated sulphuric acid produces a mixture of gases consisting of hydrogen bromide, bromine and sulphur dioxide, as depicted in the following equations.

$$NaBr(s) + H_2SO_4(aq) \longrightarrow NaHSO_4(aq) + HBr(g)$$
$$2HBr(g) + H_2SO_4(aq) \longrightarrow Br_2(g) + SO_2(g) + 2H_2O(l)$$

So hydrogen bromide and hydrogen iodide are prepared instead by the reaction of a metal bromide or iodide with concentrated phosphoric acid, a non-oxidising acid.

$$NaBr(s) + H_3PO_4(aq) \longrightarrow NaH_2PO_4(aq) + HBr(g)$$
$$NaI(s) + H_3PO_4(aq) \longrightarrow NaH_2PO_4(aq) + HI(g)$$

Properties of hydrogen halides

Hydrogen halides are colourless gases and their boiling points are relatively low except for hydrogen fluoride. Fluorine is highly electronegative and the hydrogen bonding between HF molecules in liquid hydrogen fluoride is strong and this explains its high boiling point. The boiling point increases down from HCl to HI with the increase in molecular weight (Table 9.4).

Table 9.4: Properties of hydrogen halides

	HF	HCl	HBr	HI
Molar mass (g mol^{-1})	20.01	36.46	80.9	127.9
Boiling Point (°C)	20	–85	–67	–35
Bond energy (kJ mol^{-1})	565	421	366	299
Bond length (nm)	0.092	0.127	0.141	0.161

Hydrogen halides dissolve in water to produce acids. Hydrofluoric acid is a weak acid, while the other acids are strong acids, since they ionise completely in solution.

$$HCl(g) + H_2O(l) \longrightarrow H_3O^+(aq) + Cl^-(aq)$$

Acid strength increases down the series.

$$\underrightarrow{HF < HCl < HBr < HI}_{\text{acid strength increases}}$$

One reason for the trend in increasing acid strength is the decrease in the bond dissociation energies of hydrogen halides down the series (Table 9.4).

Tutorial: helping you learn

Progress questions

9.1 Aluminium reacts with hydrochloric acid to produce aluminium chloride and hydrogen.

$$2Al(s) + 6HCl(aq) \longrightarrow 2AlCl_3(aq) + 3H_2(g)$$

(a) Write the oxidation numbers of all atoms in the reactants and products.
(b) Name the oxidant and reductant.
(c) Write partial ionic equations for the oxidation and reduction and the overall equation for the redox reaction.

9.2 (a) What happens if a piece of
(i) Al wire is placed in a solution of iron (II) sulphate.
(ii) Fe wire is placed in a solution of aluminium sulphate.
(b) Write ionic equations for any reactions that take place.

9.3 Name the naturally occurring compound of aluminium from which aluminium metal is extracted. What are the main impurities present in the ore? Explain briefly how the ore is purified.

9.4 Explain briefly the electrolytic extraction of aluminium from aluminium oxide, naming the electrodes and giving the electrolytic reactions.

9.5 Name the most reactive element in each of the following series
(a) sodium, potassium, rubidium and caesium
(b) magnesium, calcium, strontium and barium
(c) fluorine, chlorine, bromine and iodine

9.6 Give an explanation for each of the following
(a) Group I elements exhibit only +1 oxidation state in their compounds.
(b) Sodium metal is stored in liquid paraffin.
(c) When potassium bromide is warmed with concentrated sulphuric acid a brown gas is evolved whereas with phosphoric acid a colourless, fuming gas is evolved.

9.7 Why is aluminium resistant to corrosion, even though it is a reactive metal?

9.8 Write the oxidation numbers of atoms in these species
(a) HCl (b) NaClO (c) $NaClO_3$ (d) Cl_2O (e) $NaClO_2$

9.9 In which of the following cases will there be a reaction?
(a) Zn(s) added to $MgSO_4$(aq) (b) I_2(s) added to KCl(aq)
(c) Mg(s) added to HCl(aq) (d) Fe(s) added to $CuSO_4$(aq)
Write a balanced equation for any reaction that would take place.

Glossary

activated complex An intermediate formed from the reactant molecules which can change into products or reverse to reactants.

activation energy E_a The minimum amount of energy needed for the formation of an activated complex.

addition reaction A reaction in which a molecule is added to another molecule to form a single product.

aliphatic hydrocarbon An open-chained hydrocarbon.

alkali metals Group I metals of the Periodic table.

alkaline earth metals Group II metals of the Periodic table.

alkanes A homologous series of saturated hydrocarbons with general formula C_nH_{2n+2}.

alkenes A homologous series of unsaturated hydrocarbons containing a double bond in the molecule with a general formula C_nH_{2n}.

alkyl A group derived from an alkane by removing a hydrogen atom.

alkynes A homologous series of unsaturated hydrocarbons containing a triple bond in the molecule with a general formula C_nH_{2n-2}.

anion A negatively charged ion

aromatic hydrocarbon A hydrocarbon which has one or more benzene rings.

atomic mass unit, u One-twelfth the mass of one atom of ^{12}C.

atomic number, *Z* The number of protons in the nucleus of an atom.

atomic orbital The region of space where an electron is found in an atom. An *s-orbital* is spherical and a *p-orbital* has two lobes located on opposite sides of the nucleus.

***Aufbau* or building-up principle** Electrons occupy the lowest energy level, filling it before going to a higher level.

Avogadro constant, *L* The number of atoms present in 12 g of ^{12}C ($L = 6.022 \times 10^{23}$)

bond angle The angle between two bonds formed from one atom.

bond enthalpy The heat required to break the bonds in 1 mole of gaseous diatomic molecules into gaseous atoms.

bond length The distance between the centres of two atomic nuclei which form the bond.

bonding pair A shared electron pair in a covalent bond.

carbanion An ion with a negative charge on a carbon atom.

carbocation An ion with a positive charge on a carbon atom.

catalyst A substance that increases the speed of a reaction without itself being used up.

cation A positively charged ion.

chemical kinetics The study of rates of reactions.

combustion A chemical reaction in which an organic substance combines with oxygen to form carbon dioxide and water.

contact process The industrial preparation of sulphuric acid by the oxidation of sulphur dioxide to sulphur trioxide.

co-ordinate bond (dative bond) A bond between two atoms in which the electron pair is donated by one of the two atoms.

covalent bond A bond between two atoms by a shared pair of electrons.

dipole-dipole attraction The attractive forces between polar molecules.

electrolysis A process by which chemical reactions occur by the passage of electricity through an electrolyte.

electrolyte A substance that conducts electricity in the dissolved form or molten state and undergoes chemical decomposition.

electron A negatively charged subatomic particle of mass about 1/1850 u.

electron affinity The energy released when an electron is added to a gaseous atom.

electronegativity A measure of the power of an atom to attract the shared electron pair in a covalent bond.

electronic configuration The arrangement of electrons in an atom in named atomic orbitals.

electrophile (electron-liking) A molecule or a positive ion which can accept a pair of electrons from a donor to form a co-ordinate bond; a Lewis acid.

elimination A reaction in which small molecules like hydrogen halides or water are removed from a molecule. *Example,* The elimination of a molecule of water from ethanol (C_2H_5OH) gives ethene (C_2H_4).

empirical formula A formula which shows the relative numbers of atoms of each element in a compound.

endothermic reaction A reaction which is accompanied by the absorption of heat.

enthalpy, *H* A term used by chemists to express the quantity of heat transferred at constant pressure during a reaction.

enthalpy change, ΔH The difference between the enthalpies of the products and the enthalpies of the reactants.

enthalpy diagram A diagram showing the relative levels of the enthalpies of reactants and products.

exothermic reaction A reaction which gives out heat to the surroundings.

equilibrium system A system in which both forward and reverse reactions take place at the same speed.

flux A substance added during the extraction of a metal which combines with impurities to form a slag.

Gay Lussac's law When gases react with each other, the volumes of the gases reacting and their products, if gaseous, bear a simple numerical ratio to one another, if all volumes are measured at the same temperature and pressure.

Haber process The industrial preparation of ammonia from hydrogen and nitrogen.

halogens Group VII elements of the Periodic table.

halogenoalkane (alkyl halide) A compound formed by the substitution of one or more hydrogen atoms of an alkane by halogen atom/s.

Hess's law The law states that, whether a reaction takes place in one step or through a series of steps, the total heat change remains the same.

heterogeneous catalyst A catalyst which is in a different phase from the reactants. *Example:* V_2O_5 (s) in the preparation of SO_3 (g)from SO_2 (g) and O_2 (g).

heterogeneous system A system in which at least one substance is in a different phase from that of the other substances.

homogeneous catalyst A catalyst which is in the same phase as the reactants. *Example:* H_2SO_4 (aq) in the hydrolysis of methyl methanoate, $HCOOCH_3$ (aq) to methanoic acid, HCOOH (aq) and methanol, CH_3OH (aq).

homogeneous system A system in which all the reactants and products are in the same phase.

homologous series A series of compounds with similar chemical properties and gradually varying physical properties; all the members of which can be represented by a general formula, and each member in the series differs from the next by a $-CH_2-$ group.

Hund's rule of maximum multiplicity When there is more than one orbital of equal energy value, electrons occupy them singly before pairing takes place.

hydration A reaction in which water is added to a substance.

hydrogenation A reaction in which hydrogen is added to a substance.

hydrolysis The breakdown of a substance by its reaction with water.

hydrogen bond The intermolecular forces of attraction between the δ+ve hydrogen atom and δ–ve atom in compounds of hydrogen with a more electronegative element.

ionic bond The attraction between positive and negative ions.

ionic compound A compound that consists of ions.

ionisation energy The energy required to remove an electron from a gaseous atom.

isomers Compounds with the same molecular formula but different structural formulae hence different properties.

isotopes Atoms of the same element with equal number of protons but a different number of neutrons, hence different masses.

Le Chatlier's principle If a stress (such as changing concentration, pressure or temperature) is applied to a system at equilibrium, the system adjusts in a direction so as to cancel the effect of the change.

Lewis acid A molecule or positive ion which can accept a pair of electrons.

Lewis base A molecule or negative ion which can donate a pair of electrons.

lone pair An electron pair in the valence shell of an atom which does not take part in bonding.

mass number The total number of protons and neutrons in an atom.

mass spectrometry The technique to determine the masses and relative abundance of atoms and molecules.

metallic bond The attractive forces between the delocalised electron cloud and positive atomic kernels in a metal.

mineral A naturally occurring inorganic compound.

molar mass The mass of 1 mole of a substance. Molar mass is expressed in g mol^{-1}.

molar volume The volume occupied by 1 mole of a substance. The value of molar volume is constant for all gases at constant temperature and pressure and equals to 24.0 dm^3 at 1 atm. pressure and 293 K.

molarity (**molar concentration**) The number of moles of a solute present in 1 dm^3 of a solution.

molecular formula A formula which gives chemical symbols and subscripts showing the actual numbers of atoms of each element in one molecule of a substance.

nucleophile (nucleus or positive-liking) A molecule or ion which can donate a pair of electrons to form a co-ordinate bond; a Lewis base.

nucleophilic substitution A reaction in which an atom (group) in a compound is substituted by a nucleophile.

neutron A neutral subatomic particle of mass 1u.

ore A mineral from which a metal is extracted.

oxidation (1) A reaction in which a substance loses electrons. (2) A reaction in which the oxidation number of an element is increased.

oxidation number The effective number of charges on an atom in a molecule which is determined using a set of rules.

oxidising agent (**oxidant**) (1) An electron acceptor. (2) A substance that decreases in oxidation number during a redox reaction.

partial pressure The pressure exerted by a gas in a container if it alone were present in the container.

Pauli exclusion principle An orbital is occupied by a maximum of two electrons and these spin in opposite directions.

Periodic table An arrangement of elements in Groups (vertical columns) and Periods (horizontal rows) based on atomic number and electronic configuration in a way to show the relationship between the properties of the elements.

polymerisation A reaction in which many small molecules called monomers bond together to form large molecules called polymers.

proton A positively charged subatomic particle of mass 1u.

reaction rate The amount of a substance reacted or formed in unit time.

redox reaction A reaction in which oxidation and reduction take place simultaneously.

reducing agent (**reductant**) (1) An electron donor. (2) A substance that increases in oxidation number during a redox reaction.

reduction (1) A reaction in which a substance gains electrons. (2) A reaction in which the oxidation number of an element is decreased.

relative atomic mass, A_r The mass of an atom of an element in relation to the mass of an atom of ^{12}C which has a mass of 12u.

relative molecular mass, M_r The mass of a molecule of a substance in relation to the mass of an atom of ^{12}C.

saturated hydrocarbon A hydrocarbon with no carbon-carbon double or triple bonds.

slag A substance formed when a flux reacts with an impurity.

standard enthalpy of combustion, ΔH^0_c, The heat evolved when 1 mole of a substance is burnt completely in oxygen at standard temperature (298 K, 25°C) and pressure (1 atm.).

standard enthalpy of formation, ΔH^0_f, The heat change when 1 mole of a substance is formed from its constituent elements at standard temperature (298 K, 25°C) and pressure (1 atm.).

standard enthalpy of neutralisation The heat evolved when 1 mole of H_2O is formed from H^+ ions and OH^- ions provided by a dilute acid and alkali respectively under standard conditions.

standard enthalpy of reaction, ΔH^0_{rxn}, The heat change when the number of moles of the reactants given by a stoichiometric equation react completely under standard conditions.

stoichiometric equation A balanced chemical equation.

structural formula A pictorial representation to show how the atoms in a molecule are bonded.

substitution (replacement) A reaction in which an atom (group) is replaced by another atom (group).

thermochemistry The study of heat changes in chemical reactions.

valence shell electron pair repulsion (VSEPR) theory A theory which states that valence shell electron pairs (both bonding and lone pairs) arrange themselves around the central atom as far apart as possible so that there is minimum electron repulsion.

van der Waals forces The attractive forces between instantaneous dipoles.

unsaturated hydrocarbon A hydrocarbon which has a carbon-carbon double or triple bond in the molecule.

Answers to Practice Questions

Chapter 1

1.1

Element	Atomic number (Z)	Mass number (A)	Number of protons	Number of electrons	Number of neutrons
(a) He	2	4	2	2	2
(b) Be	4	9	4	4	5
(c) B	5	11	5	5	6

1.2 (a) $2s$ (b) $2p$ (c) $3p$ (d) $4s$ (e) $3d$

1.3

(a) F: $1s^2 2s^2 2p^5$

(b) Al: $1s^2 2s^2 2p^6 3s^2 3p^1$

(c) K: $1s^2 2s^2 2p^6 3s^2 3p^6 4s^1$

(d) V: $1s^2 2s^2 2p^6 3s^2 3p^6 4s^2 3d^3$

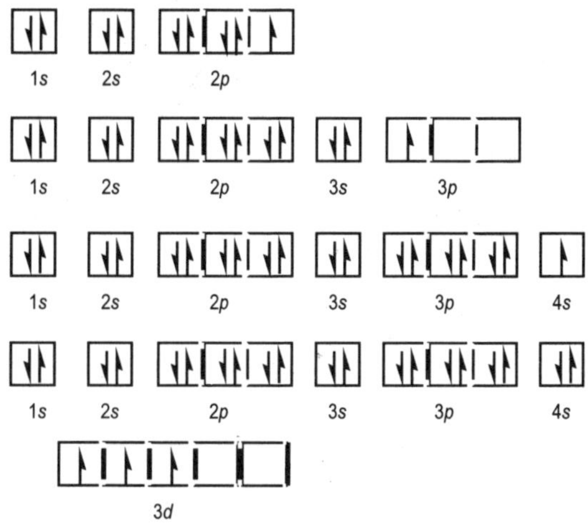

1.4

(a) Li: $1s^2 2s^1$
Na: $1s^2 2s^2 2p^6 3s^1$
K: $1s^2 2s^2 2p^6 3s^2 3p^6 4s^1$

(b) O: $1s^2 2s^2 2p^4$
S: $1s^2 2s^2 2p^6 3s^2 3p^4$
Se: $1s^2 2s^2 2p^6 3s^2 3p^6 3d^{10} 4s^2 4p^4$

Chapter 2

2.1 35.453

2.2 (a) 98.08 (b) 44.11 (c) 180.18 (d) 203.33

2.3 (a) 0.05 mol (b) 3.011×10^{22} molecules

2.4 Empirical formula = C_3H_9N, molecular formula = C_3H_9N

Chapter 3

3.1 B: $1s^2\ 2s^2 2p^1$ Al: $1s^2\ 2s^2 2p^6\ 3s^2 3p^1$

Both B and Al are in Group III of the Periodic table. It is a p^1 electron that is removed during the ionisation in both cases. Al is larger than B and the p^1 electron in Al is farther from the nucleus and so is attracted to the nucleus by a weaker force. The screening effect for the $3p^1$ electron in Al is more than for the $2p^1$ electron in B. So Al has a lower ionisation energy.

3.2 The covalent character of silver halides increases in the order shown below. As the size of the halide ion increases from Cl^- to I^-, the polarizability of the ions increases.

$$\underset{\text{covalent nature increases}}{\xrightarrow{\text{AgCl} < \text{AgBr} < \text{AgI}}}$$

3.3 As the atomic mass as well as the electron density of the noble gases increase from He to Xe, van der Waals forces of attraction between atoms increase and so the boiling point increases.

Chapter 5

5.1

(1) $$K_c = \frac{[CO][Cl_2]}{[COCl_2]} \text{ mol dm}^{-3}$$

(2) $$K_c = \frac{[SO_2]^2}{[O_2]^3} \text{ mol}^{-1} \text{ dm}^3$$

5.2

(1) $$K_p = \frac{P^2_{SO_3}}{P^2_{SO_2} \times P_{O_2}} \text{ atm}^{-1}$$

(2) $$K_p = \frac{P_{CO} \times P_{Cl_2}}{P_{COCl_2}} \text{ atm}$$

5.3

$$K_p = \frac{P^2_{NO} \times P_{O_2}}{P^2_{NO_2}} \text{ atm}$$

$$= \frac{(0.2)^2 \text{atm}^2 \times 150 \text{atm}}{(0.2)^2 \text{ atm}^2} = 150 \text{ atm}$$

5.4

$$N_2(g) + O_2(g) \rightleftharpoons 2NO(g), \ \Delta H = +90 \text{ kJ mol}^{-1}$$

$$K_c = \frac{[NO^2]}{[N_2] \ [O_2]}$$

Since the formation of NO gas is endothermic, an increase in temperature favours the production of NO gas (Equilibrium shifts to the right.). So the value of K_c is higher at a higher temperature.

Chapter 7

7.1 (a) $Na \rightarrow Na^+ + e^-$

(b) $Fe \rightarrow Fe^{2+} + 2e^-$

(c) $Fe \rightarrow Fe^{3+} + 3e^-$

(d) $Fe^{2+} \rightarrow Fe^{3+} + e^-$

(e) $Br^- \rightarrow \frac{1}{2}Br_2 + e^-$

7.2 (a) $Br_2 + 2e^- \rightarrow 2Br^-$

(b) $Fe^{3+} + e^- \rightarrow Fe^{2+}$

(c) $Cu^{2+} + 2e^- \rightarrow Cu$

(d) $H^+ + e^- \rightarrow \frac{1}{2}H_2$

(e) $Al^{3+} + 3e^- \rightarrow Al$

7.3 $Cl_2 + 2e^- \rightarrow 2Cl^-$ — reduction

$2Br^- \rightarrow Br_2 + 2e^-$ — oxidation

$Cl_2 + 2Br^- \rightarrow 2Cl^- + Br_2$ — redox

Reductant: Br^- Oxidant: Cl_2

7.4 $Ca \rightarrow Ca^{2+} + 2e^-$ — oxidation

$Br_2 + 2e^- \rightarrow 2Br^-$ –reduction

$Ca + Br_2 \rightarrow Ca^{2+} + 2Br^-$ –redox

7.5

(a) $\overset{(+1)(-1)}{Na\,Cl}(aq) + \overset{(+1)(+5)(-2)}{Ag\,N\,O_3}(aq) \rightarrow \overset{(+1)(+5)(-2)}{Na\,N\,O_3}(aq) + \overset{(+1)(-1)}{Ag\,Cl}(s)$:Not redox

(b) $\overset{(0)}{4AI}(S) + \overset{(0)}{3O_2}(g) \rightarrow \overset{(+3)(-2)}{2AI_2O_3}(s)$:Redox

(c) $\overset{(0)}{Ca}\,(s) + \overset{(+1)(-2)}{2H_2O} \rightarrow \overset{(+2)(-2)(+1)}{Ca\,(O\,H)_2}(aq) + \overset{(0)}{H_2}(g)$:Redox

(d) $\overset{(+1)(+1)(+4)(-2)}{K\,H\,C\,O_3}(s) + \overset{(+1)(-1)}{HCI}(aq) \rightarrow \overset{(+1)(-1)}{KCl}(aq) + \overset{(+4)(-2)}{C\,O_2}\,(g) + \overset{(+1)(-2)}{H_2O}(1)$:Not redox

Chapter 8

8.1 (a) CH_2O (b) CH_2 (c) CuI (d) CH_5N (e) C_2H_5 (f) CH_2

8.2

$$2C_3H_6(g) + 9O_2(g) \rightarrow 6CO_2(g) + 6H_2O\ (l)$$
$$C_4H_8(g) + 6O_2(g) \rightarrow 4CO_2(g) + 4H_2O\ (l)$$

8.3

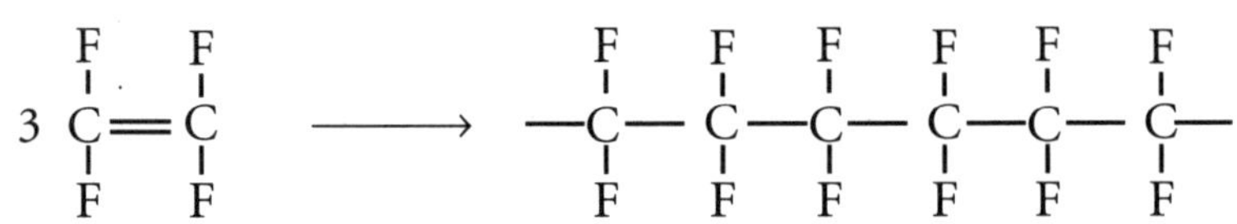

Tetrafluoroethene Teflon

8.4 (a) Iodoethane (primary)
(b) 2-Bromobutane (secondary)
(c) 2-Iodo-2-methylpropane (tertiary)
(d) 1-Chloro-2,2-dimethylpropane (primary)

8.5 (a) 2-Methylbutan-2-ol
(b) 2-Chloropropan-2-ol
(c) 3,3-Dimethylbutan-1-ol

Chapter 9

9.1 (a) Na: $1s^2\ 2s^2 2p^6\ 3s^1$
Na+: $1s^2\ 2s^2 2p^6$

(b) The first ionisation energy of Na is low. This corresponds to the removal of a single *s* electron from the third energy level of Na. The second ionisation energy of Na is very high. Na^+ is stable with the electronic configuration of a noble gas. The second ionisation requires the removal of an electron from a *p* orbital of the second energy level. This electron is closer to the nucleus and is attracted to the nucleus by a greater force. It is more difficult to remove a 2*p* electron from a filled orbit than a single 3*s* electron.

(c) No, it requires a large amount of energy to form Na^{2+} ion. This energy will not be available during chemical reactions.

9.2 (a) $2K + 2H_2O \rightarrow 2KOH + H_2$

(b) $K_2O + H_2O \rightarrow 2KOH$

(c) $2K + Br_2 \rightarrow 2KBr$

(d) $2K + S \rightarrow K_2S$

9.3 $CaBr_2$: Ca^{2+} and Br^- $\quad$ CaBr : Ca^+ and Br^-

Ca^{2+} : $1s^2\ 2s^22p^6\ 3s^23p^6$ $\quad$ Ca^+ : $1s^2\ 2s^22p^6\ 3s^23p^6\ 4s^1$

Ca^{2+} is stable with a noble gas electron configuration. Though the formation of Ca^{2+} requires a large amount of energy, this will be available during the reaction from the formation of the lattice between doubly positive Ca^{2+} and Br^- ions. Ca^+ is not stable and the energy that would evolve during the lattice formation of CaBr from Ca^+ and Br^- is small.

9.4

$$Cl_2 + 2e^- \rightarrow 2Cl^- \quad \text{— Reduction}$$

$$Cl_2 + 4OH^- \rightarrow 2ClO^- + 2H_2O + 2e^- \quad \text{— Oxidation}$$

$$2Cl_2 + 4OH^- \rightarrow 2Cl^- + 2ClO^- + 2H_2O \quad \text{— Redox}$$

$$\text{Or, } Cl_2 + 2OH^- \rightarrow Cl^- + ClO^- + H_2O \quad \text{— Redox}$$

Answers to Selected Progress Questions

Chapter 1

1.3 Isotopes: (b) and (e), (c) and (d)

1.4 $^{234}U_{92}$: 92 protons, 92 electrons and 142 neutrons

$^{235}U_{92}$: 92 protons, 92 electrons and 143 neutrons

$^{238}U_{92}$: 92 protons, 92 electrons and 146 neutrons

Chapter 2

2.1 238.030

2.2 (a) 2.0 mol He (b) 5.0×10^{-4} mol $CaCO_3$
(c) 16.6 mol Fe (d) 1.5×10^{-2} mol CO_2
(e) 0.23 mol NO

2.3 Order of increasing number of molecules:
0.25 g He < 5.5 g CO_2 < 10.0 g N_2 < 50.0 g H_2SO_4 < 2.5 g H_2 < 80.0 g C_2H_6

2.4 Order of increasing number of molecules:
3.0×10^{20} H_2O molecules < 5.0 dm^3 CO_2 < 9.6 dm^3 SO_2 < 20.0 g NH_3

2.5 Empirical formula: CH_2 Molecular formula: C_4H_8

2.6 490.0 g H_2SO_4

2.7 (a) 0.34 mol dm^{-3} (b) 0.625 mol dm^{-3}

(c) 0.10 mol dm^{-3} (d) 0.04 mol dm^{-3}

2.8 (a) 0.2 mol HCl (b) 0.1 mol Mg (c) 2.4 g Mg

2.9 (a) 9 dm^3 H_2 (b) 6 dm^3 NH_3

Chapter 3

3.9 (a) bent (b) linear (c) triangular planar
(d) octahedral (e) tetrahedral (f) trigonal pyramidal

3.10 Ionic compounds: $NaNO_2$, CaO, $MgBr_2$, $CuSO_4$
Covalent compounds: NO_2, PCl_5, SiO_2, ClO_2

3.12 Order of increasing polarity:
(a) AgI < AgBr < AgCl
(b) HBr < HCl < HF
(c) BeO < MgO < CaO

Chapter 4

4.1 –17.72 kJ mol^{-1}

4.2 5.56×10^4 kJ

4.3 +1.9 kJ mol^{-1}

4.5 –398.4 kJ

4.6 –1366.7 kJ mol^{-1}

4.7 –1189.2 kJ

4.8 +49.0 kJ mol^{-1}

4.9 –92.9 kJ

Chapter 5

5.5 $K_c = 54$

Chapter 7

7.5 (i) ON of calcium in Ca is (0) and in $CaCl_2$ (+2).
ON of hydrogen in HCl is (+1) and in H_2(0).
Ca is the reductant (ON is increased).
HCl is the oxidant (ON of H is decreased).

(ii) ON of chlorine in Cl_2 is (0) and in KCl (–1).
ON of iodine in KI is (–1) and in I_2 (0).
Cl_2 is the oxidant (ON is decreased).
KI is the reductant (ON of I is increased).

(iii) No change in ON of any of the atoms. Not a redox reaction.

(iv) ON of iron in Fe is (0) and in $Fe(NO_3)_2$, (+2).
ON of silver in $AgNO_3$ is (+1) and in Ag, (0).
Fe is the reductant (ON is increased).
$AgNO_3$ is the oxidant (ON of Ag is decreased).

7.6 In this question, the oxidant and the reductant are identified in each of the cases. Assign oxidation numbers and write the equations as required in the question. Note that in (iii) Cl_2 acts as an oxidant as well as reductant.

	Oxidant	Reductant		Oxidant	Reductant
(i)	Cl_2	Fe	(v)	$FeCl_2$	Mg
(ii)	Cl_2	KI	(vi)	H_2SO_4	Ca
(iii)	Cl_2	Cl_2	(vii)	$Cr_2O_7^{2-}$	Fe^{2+}
(iv)	$AgNO_3$	Cu	(viii)	MnO_4^-	H_2O_2

Chapter 8

8.3 (a) 1-chloro-2-methylpropane
(b) 1-chloroprop-1-ene
(c) 2-chloro-2-methylpropan-1-ol
(d) 1,2-dichloroethene

8.4 Given below are the name of an isomer of each of the compounds in (a) to (d). Draw the structural formulae of these compounds together with the structural formulae of the compounds (a) to (d).

(a) 1-chloropropane
(b) 1,1-dibromopropane
(c) but-1-ene
(d) butan-1-ol

Chapter 9

9.8 Oxidation numbers (a) H : (+1), Cl : (−1)
(b) Na : (+1), Cl : (+1), O : (−2)
(c) Na : (+1), Cl : (+5), O : (−2)
(d) Cl : (+1), O : (−2)
(e) Na : (+1), Cl : (+3), O : (−2)

Index

The Periodic Table

I	II											III	IV	V	VI	VII	VIII
1 **H** 1.0																	2 **He** 4.0
3 **Li** 6.9	4 **Be** 9.0											5 **B** 10.8	6 **C** 12.0	7 **N** 14.0	8 **O** 16.0	9 **F** 19.0	10 **Ne** 20.2
11 **Na** 23.0	12 **Mg** 24.3											13 **Al** 27.0	14 **Si** 28.1	15 **P** 31.0	16 **S** 32.1	17 **Cl** 35.5	18 **Ar** 40.0
19 **K** 39.1	20 **Ca** 40.1	21 **Sc** 45.0	22 **Ti** 47.9	23 **V** 50.9	24 **Cr** 52.0	25 **Mn** 54.9	26 **Fe** 55.9	27 **Co** 58.9	28 **Ni** 58.7	29 **Cu** 63.6	30 **Zn** 65.4	31 **Ga** 69.7	32 **Ge** 72.6	33 **As** 74.9	34 **Se** 79.0	35 **Br** 79.9	36 **Kr** 83.8
37 **Rb** 85.5	38 **Sr** 87.6	39 **Y** 88.9	40 **Zr** 91.2	41 **Nb** 92.9	42 **Mo** 95.9	43 **Tc** 98.9	44 **Ru** 101.1	45 **Rh** 102.9	46 **Pd** 106.4	47 **Ag** 107.9	48 **Cd** 112.4	49 **In** 114.8	50 **Sn** 118.7	51 **Sb** 121.8	52 **Te** 127.6	53 **I** 126.9	54 **Xe** 131.3
55 **Cs** 132.9	56 **Ba** 137.3	57 **La** 138.9	72 **Hf** 178.5	73 **Ta** 181.0	74 **W** 183.9	75 **Re** 186.2	76 **Os** 190.2	77 **Ir** 192.2	78 **Pt** 195.1	79 **Au** 197.0	80 **Hg** 200.6	81 **Tl** 204.4	82 **Pb** 207.2	83 **Bi** 209.0	84 **Po** (210)	85 **At** (210)	86 **Rn** (222)
87 **Fr** (223)	88 **Ra** (226)	89 **Ac** (227)															
			58 **Ce** 140.1	59 **Pr** 140.9	60 **Nd** 144.2	61 **Pm** 146.9	62 **Sm** 150.4	63 **Eu** 152.0	64 **Gd** 157.3	65 **Tb** 158.9	66 **Dy** 162.5	67 **Ho** 164.9	68 **Er** 167.3	69 **Tm** 168.9	70 **Yb** 173.0	71 **Lu** 175.0	
			90 **Th** 232.0	91 **Pa** 231.0	92 **U** 238.0	93 **Np** (237)	94 **Pu** (244)	95 **Am** (243)	96 **Cm** (247)	97 **Bk** (247)	98 **Cf** (251)	99 **Es** (254)	100 **Fm** (257)	101 **Md** (258)	102 **No** (259)	103 **Lr** (260)	

Atomic Number
Element
Atomic Mass